On-treatment Verification Imaging

Series in Medical Physics and Biomedical Engineering

Series Editors: John G. Webster, Russell Ritenour, Slavik Tabakov and Kwan-Hoong Ng

Recent books in the series:

Proton Therapy Physics, Second Edition
Harald Paganetti (Ed)

Mixed and Augmented Reality in Medicine
Terry M. Peters, Cristian A. Linte, Ziv Yaniv, Jacqueline Williams (Eds)

Graphics Processing Unit-Based High Performance Computing in Radiation Therapy
Xun Jia, Steve B. Jiang (Eds)

Clinical Radiotherapy Physics with MATLAB: A Problem-Solving Approach
Pavel Dvorak

Advanced and Emerging Technologies in Radiation Oncology Physics
Siyong Kim, John W. Wong (Eds)

Advances in Particle Therapy: A Multidisciplinary Approach
Manjit Dosanjh, Jacques Bernier (Eds)

Radiotherapy and Clinical Radiobiology of Head and Neck Cancer
Loredana G. Marcu, Iuliana Toma-Dasu, Alexandru Dasu, Claes Mercke

Problems and Solutions in Medical Physics: Diagnostic Imaging Physics
Kwan Hoong Ng, Jeannie Hsiu Ding Wong, Geoffrey D. Clarke (Eds)

A Guide to Outcome Modelling In Radiotherapy and Oncology: Listening to the Data
Issam El Naqa (Ed)

For more information about this series, please visit: https://www.crcpress.com/Series-in-Medical-Physics-and-Biomedical-Engineering/book-series/CHMEPHBIOENG

On-treatment Verification Imaging
A Study Guide for IGRT

by

Mike Kirby

University of Liverpool, UK

Kerrie-Anne Calder

University of Liverpool, UK

CRC Press
Taylor & Francis Group
Boca Raton London New York

CRC Press is an imprint of the
Taylor & Francis Group, an **informa** business

CRC Press
Taylor & Francis Group
6000 Broken Sound Parkway NW, Suite 300
Boca Raton, FL 33487-2742

First issued in paperback 2020

ISBN-13: 978-1-138-49991-1 (hbk)
ISBN-13: 978-0-367-77990-0 (pbk)

Library of Congress Cataloging-in-Publication Data

Names: Kirby, Mike (Lecturer in radiotherapy physics), author. | Calder, Kerrie-Anne, author.
Title: On-treatment verification imaging : a study guide for IGRT / Mike Kirby and Kerrie-Anne Calder.
Other titles: Series in medical physics and biomedical engineering.
Description: Boca Raton, FL : CRC Press, Taylor & Francis Group, [2019] | Series: Series in medical physics and biomedical engineering
Identifiers: LCCN 2018060146| ISBN 9781138499911 (hbk) | ISBN 1138499919 (hbk) | ISBN 9781351007764 (ebook) | ISBN 1351007769 (ebook)
Subjects: LCSH: Image-guided radiation therapy. | Image processing—Digital techniques. | Image analysis. | Cancer—Radiotherapy.
Classification: LCC RC271.R3 K57 2019 | DDC 616.99/40642—dc23
LC record available at https://lccn.loc.gov/2018060146

Visit the Taylor & Francis Web site at
http://www.taylorandfrancis.com

and the CRC Press Web site at
http://www.crcpress.com

Contents

About the Series

The *Series in Medical Physics and Biomedical Engineering* describes the applications of physical sciences, engineering, and mathematics in medicine and clinical research.

The series seeks (but is not restricted to) publications in the following topics:

- Artificial organs
- Assistive technology
- Bioinformatics
- Bioinstrumentation
- Biomaterials
- Biomechanics
- Biomedical engineering
- Clinical engineering
- Imaging
- Implants
- Medical computing and mathematics
- Medical/surgical devices

- Patient monitoring
- Physiological measurement
- Prosthetics
- Radiation protection, health physics, and dosimetry
- Regulatory issues
- Rehabilitation engineering
- Sports medicine
- Systems physiology
- Telemedicine
- Tissue engineering
- Treatment

About the Series

The International Organization for Medical Physics

The International Organization for Medical Physics (IOMP) represents more than 18,000 medical physicists worldwide and has a membership of 80 national and 6 regional organizations, together with a number of corporate members. Individual medical physicists of all national member organisations are also automatically members.

The mission of IOMP is to advance medical physics practice worldwide by disseminating scientific and technical information, fostering the educational and professional development of medical physics, and promoting the highest quality medical physics services for patients.

A World Congress on Medical Physics and Biomedical Engineering is held every three years in cooperation with International Federation for Medical and Biological Engineering (IFMBE) and International Union for Physics and Engineering Sciences in Medicine (IUPESM). A regionally based international conference, the International Congress of Medical Physics (ICMP) is held between world congresses. IOMP also sponsors international conferences, workshops and courses.

The IOMP has several programmes to assist medical physicists in developing countries. The joint IOMP Library Programme supports 75 active libraries in 43 developing countries, and the Used Equipment Programme coordinates equipment donations. The Travel Assistance Programme provides a limited number of grants to enable physicists to attend the world congresses.

IOMP cosponsors the *Journal of Applied Clinical Medical Physics*. The IOMP publishes an electronic bulletin, *Medical Physics World*, twice a year. IOMP also publishes e-Zine, an electronic news letter about six times a year. IOMP has an agreement with Taylor & Francis for the publication of the *Medical Physics and Biomedical Engineering* series of textbooks. IOMP members receive a discount.

IOMP collaborates with international organizations, such as the World Health Organization (WHO), the International Atomic Energy Agency (IAEA) and other international professional bodies, such as the International Radiation Protection Association

(IRPA) and the International Commission on Radiological Protection (ICRP), to promote the development of medical physics and the safe use of radiation and medical devices.

Guidance on education, training, and professional development of medical physicists is issued by IOMP, which is collaborating with other professional organizations in development of a professional certification system for medical physicists that can be implemented on a global basis.

The IOMP website (www.iomp.org) contains information on all the activities of the IOMP, policy statements 1 and 2, and the "IOMP: Review and Way Forward," which outlines all the activities of IOMP and plans for the future.

Preface

IMAGE GUIDED RADIOTHERAPY (IGRT) has revolutionized external-beam radiotherapy over the last few decades, enabling the more conformal techniques of 3DCRT, IMRT, and VMAT to be delivered with greater accuracy, precision, and confidence. It's paved the way for further advancements through adaptive radiotherapy, dose guidance, and particle therapy, to name but a few. In terms of on-treatment verification imaging, its precursors were the use of portal films and electronic portal imaging devices to verify geometrically the patient set-up; the technology and the techniques for doing this have moved on and continue to be developed, all for the benefit of our patients in terms of clinical effectiveness and safety.

On-treatment verification imaging is a term that has been used over many years for the imaging techniques centred around the treatment unit, usually called the linac, although treatment delivery is achieved through many different technologies now. It is an integral part of a seminal UK guidance document called *Towards Safer Radiotherapy*, which helped focus the minds of the radiotherapy community here in the UK on the many different aspects involved in the safe and effective delivery of radiotherapy, many strands of which are covered within this textbook. Because the practice is multifaceted and multidisciplinary, it requires many different professional talents (radiographers, physicists, clinicians, engineers etc.) along the patient's pathway to make the treatment safe and effective; on-treatment verification imaging as part of IGRT plays no small part in that.

For many years, on-treatment verification imaging for IGRT has been an important aspect of the preregistration (prequalification) teaching and training for radiographers, physicists, clinicians, and engineers going into professional practice in external-beam radiotherapy because its role holds such importance. Its importance is identified as such in the national and international training programs for all the disciplines, yet there is no detailed textbook for it. Instead, one presently has to draw upon many excellent global publications and resources on IGRT primarily addressed at the qualified radiotherapy professional. We hope with this volume we will help to redress that balance.

This textbook is designed to follow the education and training needs of preregistration therapeutic radiographers who will go on to work in radiotherapy departments around the world. It should also be an ideal study aid for those training to be qualified physicists, clinicians, and engineers in radiotherapy. Broadly, the book comprises three main sections that mirror the educational needs in on-treatment imaging for the three years of an undergraduate program of study in radiotherapy. It also contains aspects applicable to both graduate and postgraduate programs in all the major disciplines involved internationally.

An important aspect of the textbook is the accompanying electronic material that will be hosted on the publisher's website or similar, and updated on a regular basis. Within the classroom, a subject like on-treatment verification imaging for IGRT cannot be successfully taught without studying images in many different forms and exercising one's mind to think through the technology, the techniques, and the challenges of modern-day radiotherapy. This electronic resource will include images, videos, documents, and media that will complement the learning and teaching from within this book, help the student develop the enquiring mind needed for current professional practice, and help to develop radiotherapy for the future.

A final word on the terms used in this book. It is written from experience of UK practice, but with a global audience in mind—the principles and points of learning are applicable in all countries that practice radiotherapy (radiation therapy), independent of their development profile. When we use the term radiographer, we mean Therapeutic Radiographer (UK) and Radiation Therapist (RTT); when we use the term physicist, we mean trainee Radiotherapy Physicist (or Clinical Scientist) (UK) and Medical Physicist; and when we use the term clinician, we mean trainee Clinical Oncologist (UK), Physician, and Radiation Oncologist.

If you are studying and training to work in radiotherapy (radiation therapy), this book is for you!

Please visit the CRC Press website (https://www.crcpress.com/9781138499911) for colour versions of figures and further electronic resources to support learning and understanding alongside this textbook.

Acknowledgements

I'd like to thank many friends, family, and colleagues who have supported me throughout this project. The endless conversations, cups of coffee, music, meals, and words/gestures of encouragement have meant so much—thank you and praise God for all of you in Liverpool, Chester, Manchester, Preston, Ribchester, Blackburn, and many other places! This was not possible without all your love and prayers. This is dedicated to Kerrie-Anne, an amazing work colleague who, in working together so well for many years inspiring and teaching our students, was the catalyst for this project in the first place; and to my family, my brother Anthony and his family, and my dearest parents Lorna and Alf—may they rest in peace. They were, and are, the inspiration for our family lives, and why we do what we do: for the benefit of our patients. Thank you all.

Mike

I would like thank my parents for their absolute, unwavering support of me in everything I have done and achieved in my life and career. The biggest thanks go to "Team Calder"; Jason, I could not do any of the things I set out to do without you by my side, thanks for all your amazing support while I have been doing this work. My girls, Eleanor and Charlotte, all my work is and will forever be for you two. Eleanor, thanks for keeping my feet on the ground and reminding me that writing a book is "nowhere near as hard as your maths homework!" Charlotte, your smiles and cuddles make the world turn, thank you for the regular supply. Thanks for the support from all friends, including the breakfast club that kept me sane (ish!), family, and colleagues. It is fair to say this book could not have been written without the help from University of Liverpool colleagues—Jo, thanks for covering for me!

Finally the fabulous Dr. Mike, thank you so much for all the encouragement, the tips, treats, and advice. Could not and would not have achieved any of this without you. What shall we do next?

Kerrie-Anne

To Jill Stief, Andy Grocott, and Sarina Gloster at Elekta, Varian Medical Systems, and Macromedics for supplying many of the excellent images used in this document. To Louise and James and the staff at CCC Liverpool (Aintree) also for their help with some of the photographs in this book. To all at Taylor & Francis, especially Rebecca, for helping, supporting, cajoling, and guiding us throughout these last 18 months! And especially, to our colleagues at the University of Liverpool for helping us with time for writing, huge support, and putting up with us in so many different ways—thank you!

Mike and Kerrie-Anne

About the Authors

Revd Dr Mike Kirby and Mrs Kerrie-Anne Calder have more than 20 years' experience of developing and practicing the original forms of on-treatment verification imaging (electronic portal imaging) in clinical practice, helping to write the current national guidance and leading its implementation in the clinic. They have both been teaching in clinical practice for more than 18 years, and over the last 6 years in the academic setting, using blended teaching and learning methods for allied health professionals.

Mike began work in the UK's National Health Service more than 30 years ago, as a Radiotherapy Physicist at the Christie Hospital, Manchester UK. He then helped set up Rosemere Cancer Centre in Preston, UK as deputy Head of Radiotherapy Physics and Consultant Clinical Scientist. His work moved back to the Christie as Head of Radiotherapy Physics for the Satellite Centres and he helped to lead their development in Oldham and Salford as part of the Christie Network. His research has primarily been in electronic portal imaging, developing clinical practice, commissioning and technical implementation of radiotherapy technology (especially electronic portal imaging and radiotherapy networks) and, more recently, teaching and learning for radiotherapy education.

Mike has graduate and postgraduate qualifications in physics and medical physics from the Universities of Durham, Birmingham, and Manchester. He has professional membership of the Institute of Physics and Engineering in Medicine (IPEM), the American Association of Physicists in Medicine (AAPM), the American Society for Radiation Oncology (ASTRO), the British Institute of Radiology (BIR), the European Society for Radiotherapy and Oncology (ESTRO), and is a Chartered Scientist (CSci) and a Fellow of the Higher Education Academy (FHEA) in the UK. Most recently, he was invited to design and deliver a week-long regional training course on quality assurance (QA) of Record and Verify (R&V) systems as an expert lecturer for the IAEA in Algiers (2016), including all aspects of integration of on-treatment radiotherapy verification into the R&V system. As well as a lecturer (radiotherapy physics) at the University of Liverpool, he is also an Honorary Lecturer (Faculty of biology, medicine, and health) at the University of Manchester as a supervisor for the national HSST Program for doctoral level training of Clinical Scientists within the UK.

Mike is also a priest in the Church of England; having trained and studied at Westcott House and the Universities of Cambridge and Cumbria, he holds graduate and postgraduate degrees in theology. His ministry has been in the Cathedrals of Blackburn, Chester, and Liverpool (Anglican) where he is presently Cathedral Chaplain.

Kerrie-Anne began work in radiotherapy at Clatterbridge Cancer Centre 20 years ago. Following a period of work in Australia, she returned to Clatterbridge where she worked as an imaging specialist radiographer, implementing and overseeing all aspects of the imaging protocol in place. She was also involved in the education of radiographers and students for all aspects of radiotherapy imaging.

Kerrie-Anne currently works at University of Liverpool, where she educates undergraduate and postgraduate students in many aspects of radiotherapy with a special interest and role in imaging training.

Kerrie-Anne has graduate qualifications in radiotherapy and oncology from Universities of Derby and Sheffield Hallam, as well as postgraduate qualifications in teaching from University of Liverpool.

I

The Basic Foundations
in Clinical Practice

I

The Concepts and Consequences of Set-up Errors

1.1 INTRODUCTION

This initial section will introduce and discuss what is meant by the term "set-up error." The classifications of these errors will be looked at alongside the consequences that can result from the errors if the patient is treated without correction. Equipment used for error detection and prevention will be discussed in more detail in following chapters. It is important before progressing with this topic to have a full understanding of set-up errors and consequences.

Below are some important abbreviations to understand for this chapter and beyond:

GTV: Gross Tumour Volume, includes visible or measurable tumour extent

CTV: Clinical Target Volume, surrounds the GTV to give a margin for microscopic, immeasurable spread

PTV: Planning Target Volume, surrounds CTV allowing for potential errors in treatment placement.

(ICRU 2010)

1.2 DEFINITION OF "SET-UP ERROR" AND HOW THESE ERRORS ARE CLASSIFIED

In radiotherapy terms, a treatment error can be defined as any deviation in treatment delivery from what was planned/intended to be delivered (Goyal and Kataria 2014, van Herk 2004). This leads to the term "set-up error" being defined as a discrepancy between positon during treatment and planned/intended position for treatment established at planning stage (RCR 2008a Hurkmans et al 2001). Any occurrence that results in an error of treatment placement, either due to patient position or beam placement can be classed as a set-up error.

Set-up errors can occur due to a number of events and are quite commonly due to a combination of small errors or omissions. For example, an error in calculating an isocentre shift or the omission of an isocentre shift will lead to the misplacement of that isocentre position.

Some other factors that can add to or cause a set-up error are listed:

- Scanner geometry differing from linac

- Plan transfer, errors in transfer process (i.e. wrong plan attached to patient details)

- Laser position differing between pieces of equipment

- Radiographer's human error

- Patients; moving, not understanding or complying with instructions

- Organ motion, normal, internal movements

Most set-up errors occur at "expected" points in the patient set-up procedure. The use of the term "expected" may be a surprising concept: if we expect an error then surely we can remove it. The above list covers areas where we expect the possibility of a discrepancy occurring that could potentially cause a set-up error. Due to these points being "expected," we have work systems to minimise the chance of error: scanner geometry is checked at commissioning stage to ensure parity with linac geometry, plan transfer between computer planning systems, linacs and imaging systems are also investigated during commissioning, and position of lasers must be identical in pretreatment scanners as it is on all linacs and treatment equipment. Radiographers and patients are subject to human error. As health care professionals, staff are capable of introducing an error at any point in the patient's treatment; this is minimised by robust work instructions and policies that must be followed and also by working in teams for cross-checking purposes. Patient errors can be caused by patients moving, purposefully or otherwise, which can be minimised by using immobilisation, indexing the immobilisation equipment to certain points on the treatment couch, and effectively communicating with the patient about the need to remain still throughout the entire length of treatment delivery. Although every attempt is made to remove or reduce the occurrence of set-up errors, we cannot guarantee their complete removal. On-treatment imaging of the patient—in the treatment position—is vital in discovering and correcting these errors. Set-up errors can be classified as either random or systematic. Classification of set-up errors is essential to take appropriate remedial action; different actions are required depending on if the error is found to be random or systematic.

1.2.1 Systematic and Random Errors

A systematic error is defined as any error that occurs in the same direction and magnitude for each fraction (treatment) throughout the treatment course. A random error is a deviation that differs in direction and magnitude for each treatment fraction (RCR 2008a). Systematic errors can also be described as treatment preparation errors, as they are usually due to a discrepancy at treatment planning or preparation stage: incorrect instructions given, incorrect markings on patient, or misaligned lasers. Random errors can be described as treatment execution errors, as the error generally occurs during the delivery of a particular treatment, patient or organ motion, incorrect calculation of moves, or wrong immobilisation used (Gluhchev 2002). One example of an error could be a patient sneezing during treatment; this would cause the patient to move at a particular point in treatment delivery. Because this error could not be replicated in exactly the same way in subsequent treatment fractions (the patient could not sneeze at exactly the same point in treatment and move the same direction and by the same amount for every treatment delivered), this error would therefore be classified as a random error as it would be in a different direction and of a different magnitude every treatment. In contrast, if a radiographer or treating member of staff follows incorrect set-up instructions while setting the patient in position, the error is likely to be repeated every fraction, resulting in the wrong treatment position (i.e. wrong isocentre move). This would be classified as a systematic error, as it would result in the same incorrect set-up being used for every treatment—the incorrect move would be the same direction and magnitude for every treatment delivered if the error is not discovered. Systematic errors can occur to an individual patient or to a certain group of patients; for example, wrong set-up instructions may be given to one patient and cause an individual systematic error, but incorrect laser positioning in a pretreatment scanner would cause incorrect positioning for all patients scanned on that particular piece of equipment, resulting in a large group, systematic error.

1.2.2 Gross Errors

An error that potentially causes an under dose of the CTV or an unacceptably large dose to be delivered to surrounding healthy tissues outside of the PTV is considered to be a gross error. The possibility of gross errors occurring is not taken into account when creating the PTV (RCR 2008a). As a gross error could occur in any direction and at any magnitude, it is not possible to extend margins sufficiently to remove gross errors. A gross error could arise due to systematic or random errors. It is important that any gross error is detected and corrected before treatment delivery commences. The detection of a gross error can be as simple as viewing the treatment field on the surface of the patient, helped by the use of skin-rendered images produced at planning, or by acquiring an on-treatment image prior to delivery of treatment one. The size of error classified as gross will depend on the treatment margins used for each set-up. Each radiotherapy department should have clear guidance for the detection of gross errors and clear protocols on how any gross errors should be corrected.

1.3 CONSEQUENCES OF EACH TYPE OF ERROR AND THE IMPACT ON TREATMENT DELIVERED

As discussed by Van Herk (2004), a random error will cause a blurring effect on the dose distribution around the target volume, whereas a systematic error will cause a geometric shift in the cumulative dose.

The example of possible dose distribution following a number of random errors (see Figure 1.1) shows that the target area receives most of the planned treatment dose from each beam. The blurring effect is due to errors of different directions and magnitudes being present throughout the delivery of the course of treatment; this effect is not as well visualised for systematic errors.

The distribution shown in Figure 1.2 (systematic error) shows how a proportion of the target area does not receive any planned dose and an area outside of the target receives a similar, high dose from each treatment beam. If the systematic error is not corrected the dosimetric consequences will be far greater than those experienced when random errors occur.

Mathematical solutions to the issue of dose coverage have been studied and defined margins around the CTV—to account for both random and systematic errors—have been used (Stroom et al 1999, McKenzie et al 2000, Stroom and Heijmen 2002, Van Herk 2004). The mathematical formula proposed to account for these set-up uncertainties acknowledge that the deviations consist of both systematic and random errors. The models are based on the need for the majority of the CTV (>99%) to be irradiated to a high dose percentage (>95%). By applying a large PTV margin, this standard should be achieved (van Herk 2004) see Figure 1.3.

The general idea of this margin is that if we know there is likely to be a blurring effect around that target, and if we place a larger margin around the target to account for this, then, all of the target will be treated to an acceptable, clinical dose.

Applying this PTV margin to account for discrepancies in field placement is proposed to ensure CTV coverage of >99%, however, increasing the irradiated volume will lead to an increase in the amount of healthy surrounding tissue receiving a high

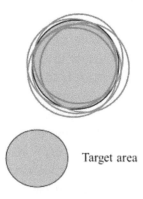

Target area

FIGURE 1.1   Example of dose distribution following random errors.

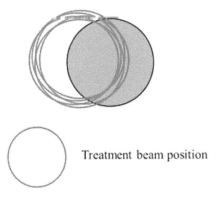

Treatment beam position

FIGURE 1.2   Example of dose distribution following systematic errors.

radiation dose. This will, in turn, limit the possibility of dose escalation to the target volume, (Witte et al 2017).

Using an enlarged PTV margin around the CTV does not guarantee the removal of all random and systematic errors. A large error—either random or systematic—could still occur and take the treatment field position outside of the enlarged PTV margin. This error would be classed as a gross error, see Figure 1.4.

Placing a large PTV around the target area may account for and minimise the impact of errors, but it will not stop errors from occurring. Errors as described earlier may still happen and cause areas outside of the enlarged PTV to be treated. It is still vital then that as many treatment errors as possible are detected and removed. The use of Image Guided Radiotherapy (IGRT) can detect many errors before they affect greatly on the dose received by the patient and can therefore be used to reduce the size of the PTV margin required (Oehler et al 2014, Gill et al 2015, Yartsev et al 2016).

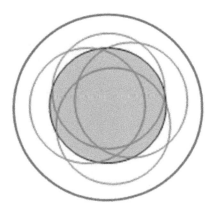

FIGURE 1.3   Example of PTV to encompass possible geometric errors.

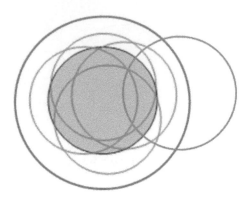

FIGURE 1.4   Example of an error occurring outside of the PTV.

1.4 CORRECTION OF SET-UP ERRORS

The consequences of each type of error must be understood to appreciate the affect each may have on the treatment delivered. An error causing a part of the PTV to be outside of the treatment field must be addressed, however, how it is resolved will greatly depend on whether that error is a random error or a systematic error. If the discrepancy is caused by a random event, instigating a permanent alteration to that patient's set-up will result in the possibility of a further error occurring, see Figure 1.5 and Figure 1.6.

Figure 1.5 shows, simply, a set-up arrangement for a simple one-field treatment. It is delivered as per plan, incorporating the whole PTV. Figure 1.6 shows a field placement error. At this point, it is not known whether it is a random or systematic error. If a permanent isocentre move is added to this treatment following this individual image, there is an equal chance that the error will be corrected or that a new set-up error will occur.

If the set-up error shown in Figure 1.6 was caused by a systematic error, the error would be removed by making a shift in the opposite direction, see Figure 1.7. If, however, the discrepancy seen in Figure 1.6 was due to a random error (i.e. one that is not likely to be repeated in subsequent fractions), a shift to correct this will result in a similar error in the opposite direction, see Figure 1.8. This is due to the original error not being present during the following fractions and the newly added shift causing a set-up error to an otherwise accurate positioning.

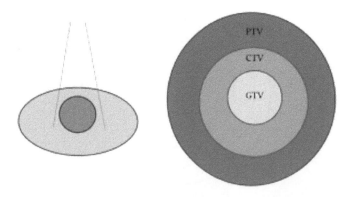

FIGURE 1.5   Treatment delivered as planned.

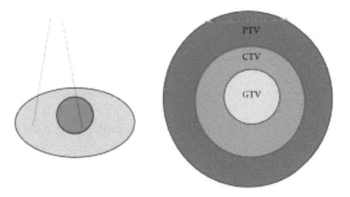

FIGURE 1.6   Treatment delivered with error.

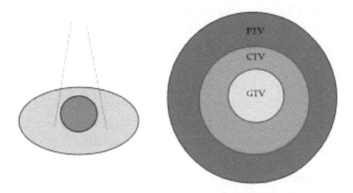

FIGURE 1.7   Error corrected following isocentre move.

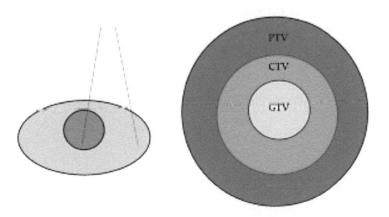

FIGURE 1.8   Error in new direction due to premature correction.

TABLE 1.1   Possible Positional Results and Effect of Applying a Permanent Shift

Original error	Subsequent error (before treatment)	Error classification	Error correction	Result
−1.5 cm	−1.5 cm	Systematic	+1.5 cm	0.0
−1.5 cm	0.0	Random	+1.5 cm	+1.5 cm

In real patient terms: A patient is set-up into the treatment position and imaged before treatment delivery. If the resulting image shows a 1.5 cm positional error (i.e. the treatment isocentre is now 1.5 cm to the left of its planned position), the treatment position will be rectified before treatment is delivered (i.e. the isocentre will be moved 1.5cm to the right). The debate now surrounds the action to be taken before the next scheduled treatment: Should the 1.5 cm right shift be applied or not? If the original 1.5 cm left error was due to a systematic error, applying the 1.5 cm right shift as a permanent instruction will remove the set-up error. However, if the original 1.5 cm left discrepancy was due to a random error that is not likely to be repeated, applying the new shift to the right will likely result in a treatment set-up error of 1.5 cm to the right, see Table 1.1.

It is important that the error classification is discovered before any decisions are made regarding altering the patients' set-up instructions. As random errors are not likely to be repeated and systematic errors are likely to be the same for every treatment, the simplest method to classify an error is to repeat the patients' set-up in exactly the same manner as planned and re-image before the next treatment is delivered. If the error remains and is similar, it can be assumed it is a systematic error, if the error is not present or is in a different direction or magnitude, a random error can be assumed, see Table 1.2.

Although all four patients in Table 1.2 have the same initial imaging results, a −1.4 cm shift, their subsequent imaging results are very different. Patients 2 and 3 have different results for the three treatment fractions, meaning the initial result of −1.4 cm was a random error. Patients 1 and 4 show very similar results that would suggest a systematic error. If a permanent correction had been applied to all four patients' treatments based solely on fraction 1, the error for patients 1 and 4 would be corrected but an extra error would have been added to the set-up for patients 2 and 3. It is essential that on-treatment images are repeated following the original set-up instructions in order to classify error type before any permanent alteration in the patient set-up occurs.

TABLE 1.2   Examples of Error Results for 4 Patients During Their 3 Treatments When Using the Same Set-up Instructions for All 3 Treatments

	Patient 1	Patient 2	Patient 3	Patient 4
#1	−1.4	−1.4	−1.4	−1.4
#2	−1.6	−0.1	+0.2	−1.3
#3	−1.3	0.0	+0.3	−0.9
Error classification	systematic	random	random	systematic

CHAPTER 2

The Foundations of Equipment Used for Radiotherapy Verification

2.1 INTRODUCTION

In this section, we will begin to look at the equipment used to ensure patients are treated in the planned position. The equipment will include pretreatment or planning equipment used to ensure accurate localisation of the target area, on-treatment imaging equipment used to ensure the target area can be suitably visualised to confirm correct positioning, and also the immobilisation equipment needed both pretreatment and during each subsequent treatment fraction. This ensures the planned position can be replicated each treatment day and also allows accurate beam placement.

2.2 IMMOBILISATION EQUIPMENT

Immobilisation is a key aspect of radiotherapy treatment delivery. The position of the patient for treatment must be achievable and reproducible for all subsequent treatment fractions. The NRIG report 2012, which looked into the use of IGRT in UK radiotherapy departments, discusses the importance of adequate immobilisation to aid reproducibility. Immobilisation is discussed before imaging systems and protocols can be decided upon. The type of immobilisation used and the patient's stability when using it will inform the imaging method, schedule, and associated tolerances. It is, therefore, an extremely important part of the patient's planning and treatment journey.

Immobilisation can involve many different concepts (see Figures 2.1–2.9). Devices can be used as an aid to position the patient; if the patient is comfortable, they are more likely to remain still and in the treatment position. Equipment may be as simple as a knee support (Figure 2.1) commonly used with a foot rest (Figure 2.2) or a simple headrest (Figure 2.8).

Other devices may be used to restrict patient movement—this is the truer meaning of the term immobilisation, as the patient is either unable to move or their movement is restricted. Examples of this immobilisation would include a thermoplastic mask used for head and neck treatments (Figure 2.6 and Figure 2.7), a thorax treatment board used for chest/breast treatments (Figure 2.3) or a prone support/belly board (Figure 2.5) (Metcalf et al 2007). Further immobilisation equipment can be used to restrict internal organ motion; these would include rectal balloons (McGary et al 2002, Cho et al 2009) and breathing belts (van Gelder et al 2018, Lovelock et al 2014). The more specialised immobilisation equipment includes devices used for stereotactic body or brain treatments. As the treatment target tends to be small, with either very small or no planning margins applied, it is crucial that the immobilisation used allows as close to no patient movement as possible. Many frames used for stereotactic radiosurgery to the brain are surgically fixed to the skull to ensure no patient movement is possible (Metcalfe et al 2007).

Examples of commonly used external immobilisation equipment are discussed here.

FIGURE 2.1 Knee support, used for comfort or immobilisation in conjunction with feet support for pelvis patients.

Source: © MacroMedics BV.

FIGURE 2.2 Feet support, used to aid reproducibility of pelvis positioning.

Source: © MacroMedics BV.

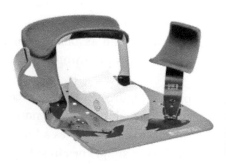

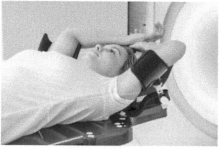

FIGURE 2.3 Thorax support, used to support and immobilise patients receiving treatments to thorax.

Source: © MacroMedics BV.

FIGURE 2.4   Omniboard, used with various accessories attached to immobilise all parts of body.

Source: © MacroMedics BV.

FIGURE 2.5   Prone head support and bellboard, used to immobilise patients treated in prone position.

Source: © MacroMedics BV.

FIGURE 2.6   3-point mask, used for immobilisation of head only, mainly used for brain treatments.

Source: © MacroMedics BV.

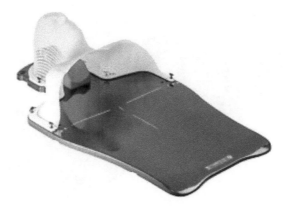

FIGURE 2.7   5-point mask, used for immobilisation of head, neck, and shoulders, mainly used for treatment to neck area.

Source: © MacroMedics BV.

FIGURE 2.8   Head rests, used in conjunction with immobilisation masks.

Source: © MacroMedics BV.

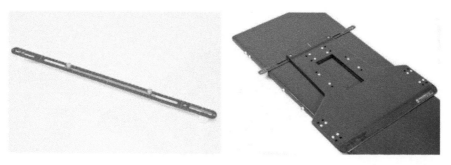

FIGURE 2.9   Indexing locator bar, used to position immobilisation equipment on to the treatment couch in a reproducible position.

Source: © MacroMedics BV.

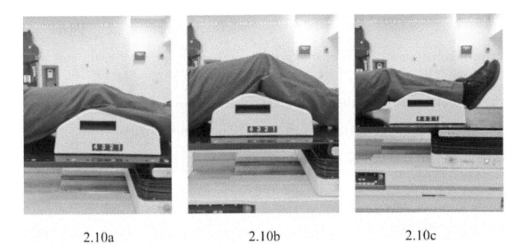

2.10a   2.10b   2.10c

FIGURE 2.10 Figures showing correct position of knee support (2.10a), knee support too high (2.10b), and knee support placed too low (2.10c).

2.2.1 Indexing

It is advised to use indexing to locate immobilisation equipment to the same place on the treatment couch for each fraction (NRIG 2012). This equipment ensures the immobilisation devices, and therefore the patient, are at a similar point on the treatment couch for every radiotherapy treatment delivered. The main aim of this is to reduce the occurrence of random set-up errors and improve reproducibility of treatment position throughout the course of radiotherapy (McCullough et al 2015, Yu et al 2017). If the patient's position on the treatment couch can be controlled, the predicted couch position technology can be used (Hadley et al 2009). Predicted couch positioning is the ability to preprogramme treatment couch parameters into the record and verify system in use. This ensures a warning is given to treating staff should the bed position differ from that acquired at initial patient set-up.

Indexing systems usually involve a simple bar located to a specific labelled (numbered or lettered) notch on the treatment couch, see Figure 2.9.

With the use of this equipment, we can ensure the patient is in a similar position for treatment every time as well as ensure the immobilisation devices are in the correct position (i.e. the patient knee support is indeed under the patient's knee (Figure 2.10a)). This sounds like a simple question that can be asked of the patient; however, it is easy for a knee support to be placed differently under a patient's thigh, or calf (Figures 2.10b and 2.10c) resulting in a very different pelvis tilt for that treatment and a possible random error. Indexing the equipment will help to remove such simple misalignments.

2.3 INTRODUCTION TO IMAGING SCIENCE

Many issues affect imaging quality that need to be considered when deciding on imaging equipment for both pretreatment/planning imaging and on-treatment/verification imaging.

These imaging characteristics include:

Scatter: Radiation arising from interactions between the primary beam and atoms in the object being imaged

Spatial resolution: The smallest object that can be seen in an image

Contrast: The ability to distinguish an object from its surroundings

Noise: Random fluctuations on the image that obscure or do not contain useful data, appears grainy

Signal-to-noise ratio: Comparison of useful signal information to background (not useful) noise

Imaging science and the associated image characteristics will be discussed more in-depth in chapter 7.

2.4 PRETREATMENT IMAGING

2.4.1 CT

Pretreatment equipment has evolved alongside treatment equipment and treatment techniques. The amount of detailed information required from pretreatment imaging when deciding upon a treatment technique has rapidly increased. There are few treatment techniques used that can be planned using traditional 2-D, orthogonal images. All techniques require volumetric imaging information and, therefore, there is a need for the patient to be scanned routinely using computed tomography (CT) scanners. CT has become the standard modality used in pretreatment imaging (Aird and Conway 2002).

2.4.1.1 Considerations When Using CT

Image quality is a main consideration when deciding modality for radiotherapy planning. A 2-D image will superimpose structures on top of each other, giving no distinction of which structure is more anterior or posterior in the patient. There is no three-dimensional aspect to the image, in a similar way that a normal photograph gives no depth perception. A volumetric image or 3-D image does not superimpose structures, rather it shows a "slice" through the patient. There is a definite depth perspective to the image as all structures and their position within the imaged volume can be clearly identified as well as their proximity and relationship to surrounding structures. The word tomography is derived from the Greek word *tomos,* meaning section, and *grapho,* meaning to write or draw. Therefore, tomography is to write or draw a section.

The machine specifications must be considered to ensure that the patient's position can be stable and reproducible for future treatments. To ensure this, the machine bore must be wide enough to accommodate immobilisation equipment. If scans are to be used for

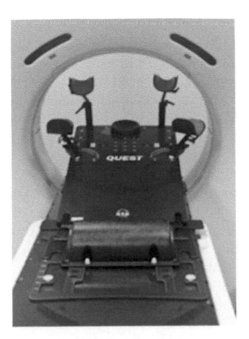

FIGURE 2.11   A chest board fitting through a wide-bore CT scanner.

planning radiotherapy treatment, the patient must be in the same position for the scan as they will be for subsequent treatment delivery. This means the scanner must be able to accommodate the immobilisation equipment to be used, especially an issue for equipment such as a chest board, see Figure 2.11.

As the patient is required to be in the same position for planning scan and treatment, there are considerations for the CT couch top. It is important that any couch deflection (couch sag) is similar to that experienced on the linac couch (NRIG 2012). Consider how much the bed might be deflected when a patient's weight is on it, and check if the bed height differs greatly when empty versus when a patient is lying on it. Also, check if any deflection experienced is mirrored by the treatment couch.

As discussed earlier in this section, indexing of immobilisation is an important issue when setting a patient into their treatment position every day. As their position on the scanner bed should be as similar as possible to that on the treatment bed, it is important to have indexing available on the CT. This ensures that all immobilisation is located at the same point on the scanner bed as it will be on the linac bed for all treatments, hopefully reducing the risk of misplaced equipment and random errors that may be caused due to patient positioning.

An additional advantage of the planning process is being able to assess the patient's ability to maintain the treatment position. CT scans are considered to be a quick imaging technique, with the actual scan time taking as little as 15 minutes. More time can be spent ensuring suitable immobilisation equipment that the patient can lie on whilst remaining still and stable for the duration of this scan as well as future treatments.

The use of a CT scanner in radiotherapy planning is vital for imaging and visualisation of target areas, including surrounding healthy tissue, and also provides dose information required for the planning process. During the CT scan, the radiation passes directly through the patient and is received by a panel detector on exit. The amount of radiation exiting the patient gives information as to how the radiation was affected by the patient's organs and tissues; this gives us essential information for the planning stage to determine the dose prescription. This will be discussed further in chapter 7.

2.4.2 MRI

Often we combine CT information with information from other imaging modalities to give a clearer indication of target areas to treat and organs to avoid. Another modality for pretreatment imaging is the magnetic resonance (MR) scan. MR scans are known to have greater soft-tissue contrast compared to CT scans. This gives a greater distinction between different organs and tissues within the scanned area. The superior soft-tissue contrast is especially useful when planning treatment to structures such as brain and spine, as there are many critical structures in these areas and visualising the exact position of each is essential. It is also useful for organs that require visualisation of a definitive boundary. For example, when visualising the boundary of the prostate in relation to the bladder, it is important to see the extent of each organ in order to cover the target and minimise the dose to an organ in close proximity. These boundaries can be difficult to distinguish using CT imaging as both organs attenuate the radiation beam in a similar way.

Better soft-tissue definition is a definite advantage of MR use, however, another important benefit is that the images are produced without ionising radiation. No dose considerations have to be accounted for when scheduling MR scans. This makes the use of repeat scans more available, either at the point of diagnosis and staging, the planning stage, or after treatment has concluded to monitor a patient's response.

Of course, as with any comparison, there are limitations to the use of MR for radiotherapy planning. MR scans, as they do not use radiation, are not able to give an indication of radiation interactions within the tissues imaged, so vital planning information is not produced. Also, MR scans are known to take a much longer time than CT scans. This can lead to patients struggling to maintain the scan position (which can cause the patient to move) as well as patients breathing and other involuntary movements (such as internal bowel movements) that can cause distortion on the image produced.

With both CT and MR giving information useful for treatment planning, a common approach is for the patients to have both a CT scan and an MR scan. The information is then combined to allow the benefits of each scan to be used and for the limitation of each to be reduced by the use of additional imaging from the other modality.

Other considerations specific to MR scans include time taken to scan, patient safety, and patient position. MR scans can take considerably more time to acquire than a standard CT scan, so patients must be informed of this and be able to remain still, usually in the treatment position, for the duration of the scan. If the patient is unable to do this, an alternative imaging modality should be considered. Due to the magnetic property of the machine, it is vital that all patients scheduled for an MR scan are thoroughly screened for any metal implants or fragments they may have in their bodies. Any patient with any metallic implant that is not MR safe must not be scanned. If the MR scan is to be used for planning purposes and the patient is required to be in the treatment position, all equipment used must be MR safe and must be able to fit through the bore of the scanner.

2.4.3 PET, PET/CT

The addition of positron emission tomography (PET) adds functional information to the imaging scans. Due to the nature of the PET scan and how the image is produced, information regarding the function of an organ or tissues can be gained. This allows radiotherapy planners to decide whether to deliver treatment to an entire mass or organ, or for treatment to be given to certain areas of it, reducing radiation delivery to healthy tissue and possible dose escalation to the target area (Erdi et al 2002, de Ruysscher et al 2005). As the PET scan gives primarily functional information, it does not give very clear subject contrast. This results in difficulty localising the exact position, shape, and size of any tissue or organ. PET scans are routinely combined with CT scans to allow both functional and structural details to be visible. PET can also be merged with MR scans.

2.4.4 Simulator

The simulator was once common, but has since declined in use at modern radiotherapy centres. The simulator was a diagnostic kV tube housed in a gantry and designed to have very similar movements and abilities as a treatment linac. The patient would be simulated in a position to be replicated during treatment. The simulator would be capable of acquiring orthogonal kV images, which would then be used to produce basic treatment plans. A separate outline of the patient would be acquired using other methods. The images produced would not show soft-tissue detail as they were not volumetric images. This is the main reason for the decline in use of the simulator. Modern, conformal treatments (IMRT, VMAT, etc.) could not be planned using a simulator. Some departments do continue to use this equipment, either due to economic reasons or for palliative treatments where it has been decided volumetric imaging is not necessary.

2.4.5 Reference Images

During the planning process, regardless of what imaging equipment is used, a reference image must be produced. This reference image will be sent, along with the plan and any instructions relating to the patient's treatment, to the treating linac. The reference image will be used as a comparison for any imaging acquired during the patient's treatment.

Any images taken during treatment will ask the question, "is this treatment in the correct position?" It is only after comparison with the reference image that an answer can be given. The reference image is a direct indication of where the treatment should be delivered. It is produced during the planning of a patient's treatment and should always match exactly the plan that has been produced.

2.5 TREATMENT DELIVERY AND ON-TREATMENT VERIFICATION IMAGING

Verification of treatment position can be achieved using a number of different techniques and pieces of equipment. The most important aspect of this verification is that the patient is in the exact position in which they will be treated. It is important that the image is taken and reviewed in a timely manner so that the patient does not have to maintain the treatment position for too long. If the wait for the patient is too long, this will increase their risk of moving (either voluntarily, for example, by adjusting position due to discomfort, or involuntarily, for example, by a cough, internal bowel movement, or sneeze). Any on-treatment verification equipment must be inside the treatment room and be able to work alongside or around the treatment delivery machine.

2.5.1 MV

Megavoltage imaging use in on-treatment verification involves the use of MV radiation from the linac head. The imaging panel, used to detect the radiation emitted and produce the image, is located directly opposite the treatment delivery head. (see Figure 2.12a and Figure 2.12b)

The MV panel can be used to acquire different types of images. The first is a direct image of the treatment beam to be delivered. As the image panel is directly opposite the treatment head, images of the exact beam shape, including MLC position, can be acquired. This gives treatment staff and clinicians the ability to see exactly what anatomy will be irradiated by the treatment beam. MV imaging is the only on-treatment imaging equipment able to provide this information. As the treatment field is being used to acquire the image, no extra dose is delivered to the patient. The dose used for imaging purposes is of the same energy and field size as the treatment field, therefore, the dose used for the image can be taken from the treatment field dose. This means the

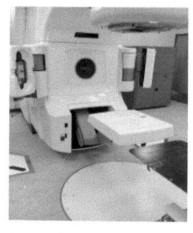
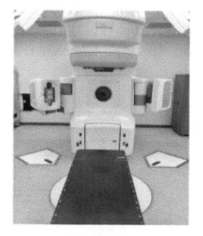

2.12a   2.12b

FIGURE 2.12   The MV panel extended and ready for use (2.12a) and The MV panel in the stored position (2.12b).

patient will only receive the dose prescribed and planned for that particular treatment field. This is a definite advantage of this type of imaging system.

The limitation when using MV imaging for treatment position verification is mainly due to the clarity of the images. If the treatment field is small, as it often is for most modern, conformal treatments, it can be too small to clearly visualise the anatomy. In this case, the treatment field must be enlarged to include more anatomy. This removes the opportunity to visualise MLC position and the exact anatomy within the treatment field. This is another type of image produced: An image that gives only a verification of the isocentre used for treatment. As the field size has been increased and the MV beam is still being used, this results in an extra dose of radiation being delivered to the patient. This MV dose can no longer be taken from the treatment beam dose as it is not of the same geometry as the planned treatment field. This results in quite a large imaging dose delivered to the patient, both in the target area and to surrounding healthy tissue as the field size is increased.

Another very important limitation of using the treatment MV beam for verification is the quality of the image produced. Due to the different photon interactions at MV energies compared to those taken using kV energies, the quality of the image produced using the MV beam does not give as clear an image as kV imaging. This can again result in a higher radiation dose delivered to the patient whilst producing an on-treatment image that does not give sufficient clarity to enable treatment staff to make a decision on treatment placement suitability.

Image quality is discussed in more depth in chapter 7.

2.5.2 kV

Kilovoltage (kV) imaging involves the addition of a kV radiation tube and detector panel being added to the standard linac design. The kV tube and detector are mounted at 90 degrees to the linac treatment head and are in addition to the MV detector panel. See Figure 2.13.

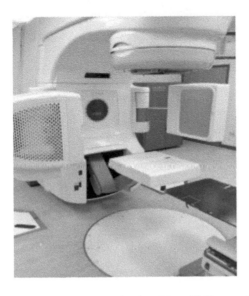

FIGURE 2.13 All the imaging equipment, both MV and kV, extended and ready to use.

The addition of this equipment produces on-treatment images of a much higher quality than MV images. Images produced are similar in quality to those used for diagnostic purposes. As the kV beam cannot replicate the treatment field, the imaging dose delivered will always be in addition to the planned treatment dose. Due to this imaging technique using kV energies, the dose delivered to the patient is quite small and a lot lower than the dose delivered when using MV energies to acquire large isocentre check fields. The quality of the images produced using the kV system has made image guided radiotherapy (IGRT) more accurate and accessible for all patients.

Although the kV tube being mounted to the gantry is the most common equipment arrangement, other systems available have the kV tube in the floor close to the linac, with the detector panels ceiling-mounted, see Figure 2.14 This gives oblique images through the patient, still at 90 degree (orthogonal) angles to allow for accurate analysis of moves required in three planes. The benefit of this system not being gantry-mounted is the position of the gantry does not affect the image orientation; the systems are separate, meaning an image can be taken during treatment (including arc therapy). Another benefit is that the position of the floor rotation does not affect the image; as long as the treatment couch does not obscure the in-floor kV tube, the treatment couch can be set to any angle and an image can still be taken. This results in images taken in the true treatment position, instead of having to move the gantry and couch rotation to the correct position for treatment following image approval.

2.5.3 CBCT

Cone beam computed tomography (CBCT) is used to produce 3-D, volumetric imaging using the treatment linac. The resulting image is similar in appearance and quality to a diagnostic CT scan, but does have some distinct differences. The main difference being the method of acquisition. In a dedicated modern CT scanner, there is one kV radiation source

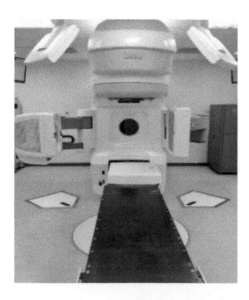

FIGURE 2.14   In-room kV tubes and detectors.

and a bank of detectors. Both parts are mounted in a thin ring aperture that the patient passes through. The ring aperture will rotate quickly around the patient as the patient is moved slowly through the scanner. This means the scan length can be taken to include the whole of the patient, if necessary, simply by setting the scan length to cover all of the patient and moving the couch through the CT scanner until all of the patient is scanned. The radiation is emitted in one thin beam that passes through the patient to the detectors as the radiation tube and the detectors rotate; the whole of the volume is imaged quite quickly. When taking a CBCT, it is not possible to have a bank of detectors attached to the side of a linac treatment machine, nor is it possible to rotate a linac at speed around the patient. No modern linac is capable of rotating more than approximately 360 degrees—one full revolution is possible, but no further. The beam shape used to acquire the scan image is also very different. For a CT scan, the kV beam produced is one single pencil beam, however, for a CBCT, the beam is a fan or "cone" shape, hence the name "cone beam CT." See Figure 2.15.

The cone shape of the beam is required due to the inability of the linac to perform continuous rotations. The CBCT has a maximum scan length available, which is the maximum field size that can be produced by the kV tube mounted on the gantry. This limits the anatomy that can be viewed, which can be an issue if treating a volume larger than the kV maximum field size.

CBCT scans are generally taken using a kV tube, but some treatment linacs have been developed using the MV beam from the treatment head to acquire the CBCT image. The same issue with maximum field length occurs whether the CBCT is taken using kV or MV equipment.

2.5.4 In-room CT

"In-room CT" or "CT on rails" are terms used to describe the addition of a diagnostic CT scanner in the radiotherapy treatment room. The advantage of such equipment is the ability to acquire a CT scan with no scan-length limitations. The most common arrangement for this equipment is to have a wide-bore CT scanner located at 180 degrees to the linac,

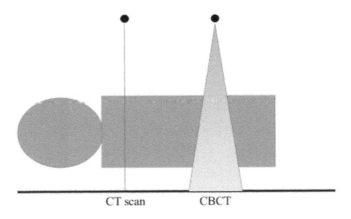

CT scan   CBCT

FIGURE 2.15   Illustration of the difference in radiation beams between CT and CBCT.

i.e. at the foot of the treatment couch. The patient is set up once on the treatment couch and two sets of lasers are generally used, one nearer to the CT scanner and a second set for the linac as standard. The CT "on rails" is then moved to allow the treatment couch to pass through its bore. Following the CT scan, the CT is moved back to its resting position and the treatment couch can be moved under the linac head, as normal, for treatment to commence. In addition to the aforementioned benefits of not having scan-length limitations, there is also the benefit of a high-quality image as the scan is taken using a full diagnostic-quality scanner. Some disadvantages do exist, however. Even though the distance between the scan and treatment positions has been minimised due to the CT scanner being able to move "on rails," there is still a chance of patient movement between the two pieces of equipment. The other main disadvantage of this arrangement is the cost of the two pieces of equipment—both to purchase and to maintain.

2.5.5 MR Linac

The use of MR scans has, until recently, been used in diagnosis only, with some exceptions in treatment planning. The use of MR images for planning has involved coregistration with a CT scan in order for electron density information to be available for the planning processes. The development of the MR linac is changing this. In a similar way that other equipment has been combined, PET/CT for example, it is hoped that the MR scanner combined with a linac will offer the best in both treatment capability and on-treatment image verification.

CHAPTER 3

Concepts of On-treatment Verification

3.1 INTRODUCTION

This section will discuss the principles and concepts involved with on-treatment geometric verification. So far, we have looked at why we need to image patients whilst on treatment, error types and the consequences associated with each type, and the equipment involved in imaging patients for radiotherapy both at the planning stage and whilst on treatment. This section will discuss the *how* and *when* of on treatment imaging.

3.2 HOW TO IMAGE

3.2.1 Reference Images

Imaging a patient to assess if they are in the correct position for treatment requires knowing what the correct position is to begin with—and we find this answer in the reference image. This image or collection of images is prepared at the same stage as the plan. After acquiring pretreatment planning scans, the images are transferred to a treatment planning system (TPS). It is within this planning system that the dose plan for treatment is produced as well as the corresponding reference image to be used during treatment.

Different types of reference images can be used. The decision of which to select will depend mainly on the treatment technique and which imaging method will allow the most appropriate anatomy structure to be visible. Another consideration is the pretreatment equipment available to produce the reference image. The reference image selected must be appropriate for verifying the patient's position during treatment and it must be able to give all information required to make a decision on the patient's position for treatment. Any reference image must be clear and must match (as near as possible) the image that will be taken during treatment, i.e. bones and soft tissue should look similar on both images. The reference image should show all appropriate and necessary anatomy within and surrounding the treatment field. It must be a high-quality image and, most importantly, it must match the treatment plan that is to be delivered. If the reference image does not match the treatment plan, we cannot guarantee the reference image will provide the correct answer to the question *"is this treatment in the right place?"* If there is a tolerance limit between the reference image and the plan, we will immediately introduce a discrepancy between the area treated and the area planned.

We will discuss this concept further when looking at tolerances for imaging and margins applied.

3.2.2 Types of Reference Images

3.2.2.1 Simulator Image

These images are produced using a treatment simulator, discussed in chapter 2. This equipment is only able to produce planar kV images, usually taken at orthogonal angles. These reference images are now mainly used for palliative treatments or other treatment techniques where it has been agreed that the patient will not benefit from a full planning scan and plan production. This is most likely to occur if the patient is unable to comply with a full planning scan. These images give the most "like for like" comparison with treatment images. As both images are produced using radiation, and neither will have been altered or enhanced using computer software, other than optimising contrast and brightness, bone and soft tissue will appear the same on both the reference and the treatment image. Simulator images are no longer used routinely in modern radiotherapy treatments due to the limitations associated with them. They cannot be used to visualise the whole treatment volume nor can they be used to visualise any soft tissue (Nagata et al 1990).

3.2.2.2 DRR—Digitally Reconstructed Radiograph

This reference image is described by its title, it is a radiograph (x-ray image) that has been reconstructed digitally, i.e. it has not been directly imaged on the patient. The patient has undergone a full CT scan of the treatment volume and surrounding organs. This CT data is then transferred to the treatment planning system for the plan to be produced. During plan production, a DRR for each treatment field will be created. This involves reconstructing the data acquired by the CT scan to produce a 2-D image of the treatment field to be delivered. Due to the whole volume being imaged at the CT scan, the DRR image can be reconstructed in any plane to match the angle of the treatment beam. The treatment field size and shape can be added to the DRR for comparison with the treatment image; this will show exactly what anatomy is to be treated. As the plan and the DRR are both produced in the TPS using the same original CT scan, there should not be an issue of the DRR and plan not matching. There will be no tolerance between field placement on the reference image and the plan produced (Dong and Boyer 1995, Van Sornsen De Koste et al 2003). When producing the DRR, all of the information contained on the CT scan is used to produce the image; no anatomy is either omitted or altered in appearance.

3.2.2.3 DCR—Digital Composite Radiograph

Composite indicates the image is made up of different parts. This image is produced in a similar way to the DRR as it uses the TPS and the full CT data. The main difference between a DRR and DCR is, during production of a DRR, all of the information contained in the CT data set is used to produce the DRR image. During production of a DCR, some of the information is altered. It maybe that some of the data on the original CT is discounted; i.e. anatomy may not be in the treatment volume and only serves to obscure visualising the treatment area. Examples of this may be omitting views of the patient's arms if obscuring a view for a chest treatment, or similarly, omitting shoulders whilst treating head and neck regions.

Other processes that can produce a DCR include enhancing certain tissue to increase visualisation on the reference image. Examples of this would include giving a higher importance to soft tissue over bone by making any soft tissue become more visible than if using all of the available anatomy. This technique can enhance bone structures also.

Care must be taken in the production of DCRs to ensure the reference image still relates to the image that will be taken whilst the patient is on treatment. As stated earlier in this chapter, a reference image must be clear and must match (as near as possible) the image that will be taken during treatment. It is very useful to be able to create a reference image without any anatomy that obscures the view of the treatment area, i.e. removing the arms or shoulders, however, this is not possible during treatment—obviously, the patient's arms cannot be removed to create a clear on-treatment image. Altering the reference image in such a way may only serve to cause confusion to staff who are evaluating the patient's position. If the DCR has been altered greatly away from the original image data acquired, consultation with staff who are to be using the reference image should be sought.

3.2.2.4 CT Data Set

The reference image used will depend on the imaging technique used for treatment. If the patient is to have 2-D verification images during treatment, a 2-D reference image is suitable. If however, the patient is to have 3-D volumetric imaging (CBCT) whilst on treatment, a 2-D reference image will not be suitable. The full CT data set, taken during the planning CT scan, will be required as the reference image. The use of the full CT data set will allow soft tissue match between the CBCT and the reference image. Any 2-D reference image will not allow for soft tissue visualisation and therefore will not allow for the CBCT to be registered. As discussed earlier in this section, any reference image must give all the information required to decide on treatment positioning and also match, as closely as possible, the image taken during treatment. Therefore, full CT data must be used as the reference image when using CBCT.

3.2.3 Verification Technique

When deciding which technique to choose for on-treatment verification, the main point to consider is: Which technique will give us the best visualisation of the treatment area? The time taken to perform the imaging and, importantly, the dose to be delivered to the patient must be considered.

There are different methods of acquiring an on-treatment verification image, each associated with a different dose and/or energy. The equipment available within a radiotherapy department will have the biggest effect on which on-treatment imaging method can be used.

3.2.3.1 MV Imaging

If the treatment volume mainly comprises bony anatomy and gives a large enough view to enable a clear anatomy match, then an MV image using the treatment field can be used. The main benefit to this imaging technique is, there will be no extra radiation given to the patient. This is because a portion of the prescribed treatment dose can be used to acquire this image. This is the only technique that allows visualisation of the exact treatment field to be delivered, including exact treatment-field size and shape. Treatment techniques that may utilise this imaging type are breast treatments or treatments to a limb, bone metastases, or sarcomas.

Should the treatment field prove too small to be used for imaging purposes, as many modern and conformal fields are, the MV field can be made larger for imaging purposes. Using this technique ensures that sufficient anatomy can be visualised in order to make a decision regarding field placement, however, it also removes the ability to visualise field size and shape. This method delivers a radiation dose to the patient—and it must be considered—as the prescribed treatment dose can no longer be used due to the different geometry of the fields. Minimise or avoid the use of this imaging technique if kV imaging is available.

3.2.3.2 kV Imaging

Kilovoltage (kV) imaging delivers a lower radiation dose to the patient than MV large fields. Although the radiation dose is still in addition to the prescribed treatment dose and, therefore, should still be considered when taking any images, the dose received by the patient is much lower compared to using a similar planar imaging technique with an MV beam. Due to the kV imaging using a kV source and panel detector in addition to the treatment head, the exact field size cannot be visualised—there are no MLCs fitted to the kV tube so it cannot replicate treatment fields. Any image taken that does not show the exact treatment-field size and shape is used as an isocentre position check only. This is not a limitation as we can assume that if the isocentre is correct, the field placement will be also. The field geometry in relation to the treatment isocentre can be checked using the planning system.

3.2.3.3 Cone Beam CT Scans (CBCT)

Cone beam CTs are the only way to take volumetric images on a standard treatment linac. CBCTs can be taken using MV or kV radiation, depending on the equipment available, but kV is the more common of the two energies used. Scans are taken using either the kV tube and detector or the treatment head and associated detector panel. The choice of energies to acquire the CBCT will be dependent on the equipment installed. The CBCT acquisition involves one full or half rotation around the patient by the treatment linac, with the patient in the treatment position. They are relatively simple to acquire and can be taken during a

TABLE 3.1 Concepts of On-treatment Verification

Imaging Technique	Equipment	Reference Image	Suitable to View	Main Considerations
MV treatment field image	MV panel and linac head	2-D image, DRR or DCR	Bony anatomy within treatment field	No extra dose delivered, cannot visualise sufficient anatomy if small field size used, shows full field size including MLC shape
MV large image	MV panel and linac head	2D image, DRR or DCR	Bony anatomy within and surrounding treatment field	Large image size used, shows anatomy outside of treatment field, large MV dose delivered
kV image	kV tube and panel	2D image, DRR or DCR	Bony anatomy within and surrounding treatment field, fiducial markers	Large image size used, shows anatomy outside of treatment field, much smaller dose delivered to patient than MV large field
CBCT kV	kV tube and panel detector	CT data set	Full volumetric anatomy bony and soft tissue	Able to visualise soft tissue in the whole treatment volume, ideal for areas when consideration to surrounding organs must be given, able to visualise out of plane rotations.
CBCT MV	MV panel and linac head	CT data set	Full volumetric anatomy, bony and soft tissue	Able to visualise soft tissue although commonly not as clear as kV CBCT due to higher energy. Higher dose delivered compared to kV CBCT due to MV energy

normal treatment appointment time. CBCTs are suitable for most anatomical areas (McBain et al 2006). The patient's position and machine clearance is the only limiting factor to acquiring these scans, if the gantry of the linac cannot safely rotate around the patient, the scan cannot be done. The dose associated with CBCT will, again, be in addition to the prescribed treatment dose and must be considered when deciding to image this way. See Table 3.1.

3.3 WHEN TO IMAGE

After the decision is made regarding the reference image and imaging technique, the next consideration is when to image. Do we need to image during the first fraction? What is it we are hoping to achieve in doing this? If a patient has planning images taken and proceeds to treatment the same day, is there a benefit to imaging that patient again only a few hours later? Radiotherapy treatment can often be a stressful process for the patient with many patients feeling nervous during their first treatment, this can often make them extremely tense while being positioned on the treatment couch. Images taken on the first day can give registration results that can be quite different to those taken on subsequent treatments (Ludbrook et al 2005, RCR 2008a). Due to this, there can be a case made for not taking images during the first treatment as the results will not be representative of the patient's future position. So should we not image patients on Day 1 of treatment, and thus save them the added radiation dose when the image results are likely to be unusable?

The answer to all this depends on whether we can assume every process during the planning stage and pretreatment imaging, including tattoo or permanent marker positioning, has been completed correctly. It's not that we assume our colleagues in other sections of radiotherapy *will* make an error, but moreso we cannot assume that they will *never* make an error. Tattoos can easily be placed in the wrong position (this can happen very easily should the patient move, even a very slight amount, between scan and tattooing), bed controls can be knocked, and laser positions used to locate and place tattoos can be different to those in the linac rooms. Transferring the plan to the treatment machine can be a potential source of error—can we guarantee every move for every patient has been sent to treatment correctly? No, of course we always must check. The image taken on the patient's first day of treatment should be used to identify any gross errors that may have occurred. During this gross error check, the radiotherapy delivery staff will be looking, primarily, for any major discrepancies between planned position and treatment position. Any gross errors discovered should be investigated before deciding on any correction. Treatment should not be delivered without the initial gross error check before fraction one (RCR 2008a).

After successful imaging before fraction one of treatment, when should we next image the patient? Patients can change during a course of radiotherapy. As most radical treatments will be delivered over a period of weeks, it is possible that a patient's size and shape may change. Medications or treatments the patient is undergoing can cause weight gain or weight loss, which will cause any external marks to move in relation to the target area. Setting the patient's position to these marks will not necessarily place the treatment centre in the planned position. Other factors, such as tumour shrinkage or progression, can make the treatment centre no longer in the best position. Treatment reactions, such as hair loss, skin reactions, and ability to maintain position due to pain, can make the patient lie differently on the treatment couch as treatment progresses. We must also be able to determine whether an error is random or systematic. All of these issues mean the on-treatment images taken must be repeated after image evaluation at treatment one.

When exactly to repeat the on-treatment images for patients will be discussed further in chapter 6 when we look at imaging protocols.

3.4 METHOD OF IMAGE MATCHING

Following image acquisition, each image must be evaluated. There are a number of ways in which on-treatment images can be assessed. The choice of match type will depend on the images taken as well as the anatomy within the target area. Match types include bony match, fiducial marker match, and soft tissue match. When deciding which match is most suitable, the anatomy, or surrogate, to be used must be visible on both the reference image and the image taken on-treatment.

3.4.1 Bony Match

A bony match is using the visible bony anatomy to match the on-treatment image to the reference image. This match is suitable if the target area consists of mainly bone or if the bone within the target can be used as a suitable surrogate for the target. All methods of on-treatment imaging will allow for a bony match to be done. Bony anatomy will be visible on 2-D planar MV and kV images and on CBCT scans.

3.4.2 Fiducial Marker Match

A fiducial marker is a radio-opaque marker that can be implanted into the patient either within or in close proximity to the target volume. These markers can then act as a surrogate for the target area to aid treatment-field positioning (Kothary et al 2009, Chung et al 2004). The material used for the fiducial marker may affect which imaging modality is able to be used. Most fiducial markers will be visible when using MV and kV planar imaging as well as CBCT. Fiducial marker matches are commonly used for prostate treatments.

3.4.3 Soft Tissue Match

This technique is an image match made using visible soft tissue. As soft tissue anatomy is only visible when full volumetric images are acquired, the only imaging method able to utilise this match is CBCT, either MV or kV. This match will be most suitable for treatment targets consisting mainly of soft tissue or when the area to be treated is in close proximity to soft tissue organs at risk. The reference image required for this match is the full CT data set.

3.5 WHEN TO CHECK THE IMAGE

Imaging systems and the images stored in them are readily accessible at the treatment console, whether they are integrated with the linac computer system or separate from it. All images are instantly available to be reviewed by the treating radiographers, so we have a decision to make: When is the most appropriate time to evaluate the image and make a decision on the patient's treatment position? In an *on-line review*, the images can be reviewed before that treatment fraction is delivered, whilst the patient is on the bed and in the treatment position. Alternatively, in an *off-line review*, images are taken and stored for review later, after treatment delivery, when the patient has left the department.

3.5.1 On-line Review

Using an on-line approach to image evaluation means that the patient is still on the treatment couch and in the treatment position whilst the image is reviewed. The main advantage to this method is that the patient's position, based on the imaging results, can be altered and corrected before any treatment is delivered. This would ensure, as much as possible, that the patient is in exactly the correct position before any treatment is started. There are, however, disadvantages to this method of image evaluation, with the main one being time pressure for staff. If the patient is on the treatment couch in the correct position for treatment, there is a real concern that the patient will not be able to hold that position for very long. The patient could become uncomfortable and consciously move, or internal movements can occur; either would mean that the image taken will no longer represent their current position. This produces a very real time constraint for the treatment staff evaluating that image. The image must be evaluated and any positional alterations decided, agreed, and implemented before the patient moves. The time constraints associated with this method of imaging review can be more of an issue when using CBCT. The CBCT itself takes longer to acquire than orthogonal images and produces a much larger volume of imaging data that must be analysed.

This can cause a rushed approach to the image evaluations and it minimises the chance to gain other opinions from staff who may be more experienced in image analysis.

3.5.2 Off-line Review

Analysing images using this method is done after the treatment fraction has been fully delivered, and this gives no opportunity for the patient's position to be corrected prior to full treatment dose delivery. After image review, if a move is deemed necessary, it can only be applied for the next fraction; the previous fraction will have been delivered with the error that was discovered at image analysis, which is a large drawback to this method of image review. However, using the off-line review technique removes the time pressure for the treating staff. As the patient has already left the department and had treatment, this leaves time to ensure the analysis is done correctly, to agree upon and settle any debates or uncertainties regarding the treatment position, and gain further opinions from other experienced staff.

3.5.3 Review Method

We have to decide which review method is most appropriate to use: On-line, so any errors discovered can be corrected before treatment delivery but analysis is performed under stressful and time constrained conditions, or off-line, where the treatment has been delivered and any errors found cannot be corrected until the following fraction but the image is evaluated in much less stressful conditions? Imagine being set a task, even a simple problem to solve, which you can do at leisure and seek help from experts before suggesting your solution, compared to the same task requiring your answer and action plan within a definite and very short time limit, in a busy, noisy working environment with multiple distractions. The first scenario when you can solve the problem at leisure and seek help is most likely to be completed more thoroughly and correctly than the task with the time constraint. This thought would lead us to believe that off-line analysis is the more appropriate of the review methods, however, the issue of the treatment already occurring is now added to our thoughts.

The issue of on-line versus off-line correction strategies is discussed by Goyal and Kataria 2014.

There is a way to reduce the issues associated with on-line review: Staff training.

If staff are trained using the imaging equipment and in how to analyse images under a time pressure, the risks of using on-line review of images should be reduced, leaving us and our patients to benefit from knowing treatment is in the correct position prior to delivery. We will discuss training issues further in the next chapter.

CHAPTER 4

Clinical Protocols and Imaging Training

4.1 INTRODUCTION

Protocols are written instructions, easy to both access and follow, that relate to a particular aspect of treatment. Protocols are an important feature in any radiotherapy department to ensure consistency of care for all patients and that standards are maintained in all areas of the department. Protocols will also ensure the safety of patients, visitors, and staff (Van der Merwe et al 2017, IAEA 2008).

An imaging protocol, therefore, should be the full instructions and guidance document containing the what, why, and how of all imaging to be performed in the radiotherapy department. There is likely to be a pretreatment imaging protocol and an on-treatment imaging protocol. This chapter will discuss the on-treatment imaging protocol.

It is important that any decisions made regarding on-treatment imaging are placed into the protocol to ensure all staff are aware of the standards expected and the agreed actions to be followed at all times.

It is vital in any radiotherapy department to ensure all staff are fully trained to a competent level to guarantee that the protocol is up-to-date, practical, accessible, and followed. All staff need to be aware of the imaging procedure to ensure a consistent approach to on-treatment imaging, regardless of the piece of equipment used to treat the patient or which staff are delivering that day's treatment. The on-treatment imaging protocol is not merely the concern of treatment staff—staff in the radiotherapy planning department need to know the decisions likely to be made by treatment staff and how those decisions will affect field placement and margins allowed. This information will guide planning radiographers in their clinical decisions. The writing, implementation, and staff competence with regards to an imaging protocol affects the whole radiotherapy department.

4.2 IMAGING PROTOCOLS

An imaging protocol should give clear guidance and indicate various aspects of the radio-therapy imaging process (Nilsson et al 2015), including:

How to image: Clear and concise instructions for how to carry out certain imaging procedures. All staff must be aware of how to complete imaging tasks, such as what modality to use, what images to take, and what action to take if the chosen modality is not available.

When to image: Instructions on when an image is indicated and taken routinely, when an image should be repeated, and when treating staff can justify taking an image outside of routine. For example, images can be repeated as necessary due to patient positioning issues, change in immobilisation equipment, discrepancies in distance readings, and other similar issues. The protocol should clearly describe those patients who require daily imaging and those who do not.

Image registration instructions: How will the images be evaluated? Instructions regarding on-line or off-line registration should indicate who is permitted to evaluate each image type. It may be that all staff of a certain grade are permitted to evaluate all images, or that staff are required to have completed various training and competency checks for each image and registration technique before evaluating those images. The decisions regarding permissions to register images should be clearly stated in the protocol.

Who can image: Guidance as to what qualifications or training packages must be completed to become competent in different aspects of on-treatment imaging. This will ensure that all treatment staff are aware of the training required and which staff members can be asked for advice or guidance when an image needs to be analysed.

Actions to be taken: A clearly stated list of actions to take after the image evaluation. Guidance on how a gross error should be corrected as well as the actions to take if an image registration is out of the tolerance stated for that particular treatment but falls short of a gross action level.

Recommendations for these aspects of an imaging protocol suggest a simple-to-follow flowchart (RCR 2008a) since it provides an easy-to-interpret system for obtaining imaging evaluation results and the actions to take following those results.

Tolerances: The acceptable tolerance level for all treatment arrangements. All staff must be aware of how large a positional error can be before a correction strategy must be applied.

Prescribing of associated imaging dose: Any radiation dose delivered to a patient must be justified and prescribed by an approved prescriber. The protocol should indicate how many images and what dose from imaging can be applied to a patient before seeking further approval and prescription for more.

Audit: All clinical protocols must be assessable for audit purposes. Every aspect of the protocol should be available for audit, and these audits must take place at regular intervals to ensure compliance and that the protocol remains appropriate. The protocol must be updated regularly as new treatment techniques or new imaging or immobilisation equipment is introduced. Regular audits ensure the protocol evolves alongside advances and improvements in treatment delivery.

The protocol must be evidence-based, practical, and effective to ensure the accuracy and suitability of the instructions. The use of a large document, written based on evidence-based facts and guidance, will not be easily followed if it is not practical to implement. Deviations can still occur even after implementation of a robust imaging protocol (Cvetkova et al 2018). A full staff-training programme will help to ensure compliance with protocols.

4.3 TRAINING

Training and competency in radiotherapy imaging has increased in importance as technology has evolved. Modern, on-treatment imaging capabilities have enabled radiotherapy departments to visualise more anatomy, and due to this advancing capability, on-treatment imaging is relied upon much more to inform treatment decisions. Decisions such as clinical margins, imaging tolerances for treatment techniques, changes to anatomy, and treatment position verification are all made using on-treatment imaging. Making such critical decisions requires that staff working in the radiotherapy department must be competent in all these areas. Robust training methods must be in place to ensure this competence and to give other staff confidence in those making the decisions.

Training and maintaining competency is an integral part of the electronic portal imaging (EPI) process. A training programme will ensure that each individual is trained to a consistent level which can reduce inter-observer variability (RCR 2008a).

There are many points to consider when deciding on a suitable imaging training process, especially in a large radiotherapy department. Factors influencing any training decisions include workload, staffing numbers, availability of trainers, as well as any other departmental issues.

4.3.1 Who to Train?

In a large radiotherapy department with multiple linacs working up to and beyond 12 hours per day, it would not be suitable to have one member of staff responsible for all imaging decisions. However, in a small department of only one or two treatment machines, having one member of staff taking on all imaging decisions might be preferable to ensure consistency of the service. The advantages of ensuring consistency must be considered alongside the pressure of only having one member of staff working to evaluate images. Departments should know who will cover the role when that member of staff is not available.

Most large departments tend to have a core team of staff trained to a higher level in imaging who are responsible for either completing or overseeing all decisions made regarding image analysis. When it comes to imaging training, it is important that all treating staff know who they should call for advice and guidance, when they are required to seek guidance, which staff members are available to give imaging advice, and that all of the imaging trained staff maintain competence in this field.

It is expected that all treating staff will be involved in the imaging of radiotherapy patients in some capacity. As imaging systems are generally integrated into the treating system, and since quite often both actions are controlled by the same computer system, treating staff should be able to take an image of the patient they are treating. All staff treating patients should have a basic level of training and understanding of the process of image acquisition. It is important that all radiotherapy treating staff know how to acquire an image and are fully aware of when an image should be taken. This should be explained in the department's imaging protocols. After image acquisition, analysis may be passed to other radiotherapy staff. For some treating staff, image acquisition will be their only involvement.

Analysing the image and making treatment decisions based on that analysis often requires further training to ensure competence; it is not likely that a department will be able to train every member of staff to the required higher level. Many departments choose a core number of staff for this training; deciding who should be on that team is the next decision to be made.

Staff responsible for making imaging decisions should be experienced treating radiographers and, if possible, have some experience in planning treatment. This will help with understanding of margins used for each treatment site and applying the corresponding correction strategies. It is easy to assume that all junior grades of staff do not possess knowledge to analyse images or that all senior staff do have the ability. This is not always the case, as some junior staff prove to excel at these tasks, while more senior staff, due to a variety of reasons, may lack in ability if they have not worked fully clinical for some time. Care should be taken when selecting staff and grade should not be the only factor used.

4.3.2 When to Train?

Imaging training incorporates many different aspects of the role and is usually delivered in stages. All staff need to be aware of the departmental imaging protocol and this is usually where imaging training begins: A general familiarisation with the protocol and working practices. Next comes the more technical training to ensure all staff are familiar with the imaging equipment. This training can be delivered very early in any staff member's role; the most established treating staff will be in a suitable position to deliver this training and guide new staff.

Following the reading of department protocols, any treating member of staff should now know when to take an image and when it should be repeated. The next step in training will be the more interactive aspect of actually acquiring the image. This will require training in the practicalities of the system used for imaging. How is the image actually acquired?

Which button is pressed and how much radiation is required? These steps, although vital, are fairly straight-forward to teach and are often now incorporated into general training for treatment delivery as it is such an integral part of the treatment role. Some radiographers will require more input in training than others. This is usually due to previous experience in using particular treatment machines and associated software.

When treating staff are familiar and comfortable with when an image is required and how to acquire that image, the next aspect of training is how to analyse that image. This phase is where the decision of who to train applies. Not all staff will be trained to analyse all images. Some images are routinely taken and require less input from treatment staff to analyse. These images would include simpler treatment techniques; for example, an anterior planar image for a parallel opposed technique. Generally, this image would not require complicated soft-tissue analysis. All treating staff members should be trained in analysis of simple, bone-matched treatment images. This will aid patient throughput on treatment machines as a higher trained member of staff will not be required to attend the treatment machine to analyse these simpler images. The image analysis and following actions taken may, however, be subject to audit later on. Initial image analysis training should be done using a specialist training database that ideally includes on-treatment images taken within the department. After the initial training using a separate database, the trainee can now move on to analysing images on the live treatment system following an off-set review protocol; this will remove any time pressure on the trainee. When competency is gained in the analysis of simple, bone-matched, planar images, training may then progress on to more complicated matches, such as planar images with small margins and nearby organs at risk, or on-treatment analysis; the image match will be the same but the trainee now has the added time pressure of the patient still being on the treatment couch in position.

The final aspect of on-treatment image training will be the more complicated soft-tissue matches (IAEA 2008). These are usually performed by more experienced treating staff, preferably with treatment planning experience or knowledge of treatment margins used in the department. This task is usually performed by a core team of staff trained in advanced image analysis; the decision of who to train will be at the discretion of the department to suit departmental needs.

4.3.3 How to Train?

When designing a training package, it is important to ensure that all staff undergoing a certain aspect of imaging training have access to the same training and are assessed using the same tool and criteria. This approach will ensure that all treating staff are confident in the ability of all imaging-trained colleagues without having to consider different capabilities. As discussed, some aspects of imaging training can easily complement and be delivered alongside general treatment and equipment training. Training to acquire images can be delivered this way. Image analysis training needs to be taught and explained fully. There are many factors involved in deciding if a patient's position is correct and the appropriate actions to take to rectify any errors. A separate training system should be used for image analysis training that is away from the normal work environment to allow the trainee to give the images full attention and concentration. The training database also allows for clinically relevant treatment images to be evaluated without fear of affecting clinical decisions. The database should have clinical images available that are relevant to the department's work i.e. images the trainee is likely to be required to review when deemed competent to do so. The training needs to be consistent for all staff and requires an answer to be established. The member of staff delivering the image analysis training will need to know if the trainee is getting answers correct—this is another major advantage of using a separate training database. The images can be loaded in to the database in advance, allowing the imaging trainer time to establish correct evaluations for each image as well as a correction strategy if any errors are found. Correct answers can be derived using either the real clinical results used for the patient, if known, or in collaboration with other staff members working in image analysis.

Training would usually begin with the simpler of images, such as planar, bony matches first until competence and confidence is gained. Following experience at this task, other and more complicated images can be included, such as soft-tissue matches with volumetric imaging. It is important to keep in mind when training staff in image analysis that experience of treating or professional grade does not automatically mean that one is competent in imaging. All staff need to be trained and given the same information and opportunity to be competent. Following successful completion of each stage of imaging training, trainees should complete a competency assessment. Although not a popular task, the assessment serves a very important purpose: It enables all staff to be confident in each other's ability to perform the necessary work. Staff who have been successful in completing the competency check will also be more confident in their own ability. Before undertaking the next step of image analysis training, and if time pressures permit, it is ideal to allow staff the time to gain experience in that aspect of image analysis, i.e. following successful completion of initial training and assessment in planar, bony matches, it would be ideal to gain experience completing these image matches on real patients in the live system before starting training with more complicated, soft-tissue matches. It may be more beneficial for a department to stop training for most treating staff at this point and allow only a core team of staff to complete the more in-depth training.

The use of an assessment for staff following each point of imaging training will obviously cause an issue regarding fails. If a staff member is unsuccessful at the assessment stage, what actions will be taken? Discuss and plan for this in advance of anyone taking the assessment. When considering a reassessment, care must be taken to ensure trainees are shown to be improving their image analysis skills and not improving in their ability to simply take and be successful in a test. Departments may consider a few options, including a period of further training, a new and independent assessment, a period without any more formal training to allow trainees to gain more experience with the imaging protocol for the department, further observation of trained staff working and completing image analysis, as well as more time to become familiar with the imaging equipment and software package. The training process and assessment should be repeated until the trainee reaches a point deemed competent in this task and the staff member is confident in their imaging ability.

Training using a separate database has many advantages, one of which is the ability to analyse images in a quiet teaching room, away from the general noise and distraction of a busy radiotherapy machine. This, however, is not how most staff work. The luxury of moving to a quiet space to evaluate every image taken is not possible in a work setting. The differences need to be addressed, usually by giving each newly image-trained member of staff some time to experience evaluating an image in the normal work setting. To fully gain competence in image evaluation, we must make sure all staff can correctly evaluate that image and make decisions informed by the patient's own plan and departmental protocols, but we also must to be able to complete these tasks in a busy work environment with multiple distractions in a timely manner, as it is likely the patient will be in the treatment position on the treatment couch.

Decision making following image analysis must also form an important part of imaging training. It is important that all staff analysing images are aware of appropriate actions to take after discovery of an error in treatment placement. Training and education for staff in the actions to take after image review can be done as a desktop package using a prepared workbook.

Only when all of the elements of image evaluation can be completed in a normal working environment with all of the usual distractions, can a treating member of staff be competent in image evaluation.

4.3.4 Does Training Need to Be Repeated?

In general, people who are trained in a task and successfully complete that task a number of times a day, every day, during their working hours, will become more experienced and competency and confidence will continue to improve. This can be the case if a task remains the same, however, radiotherapy treatments and associated imaging procedures often change. Regular training updates are required to ensure all previously competent staff retain that competency level with new techniques and equipment. Competency levels can also become affected by boredom and complacency. If you successfully complete the same task a number of times every working day, you might expect someone to become complacent at that task.

The two examples given so far assume that a treating member of staff will complete image analysis a number of times every day; this is not always the case. The role of treating radiographers is diversifying into many different areas. For example, a member of staff, after being deemed competent in image analysis, may then spend a significant time working away from a treatment machine either working with associated documentation or in the pretreatment scanning department. The competence of these staff members cannot be assumed to remain at the same level when they return to a treatment machine and begin to analyse on-treatment images once more.

For all of these reasons and others, competency must be re-evaluated at regular intervals. The length of these intervals will depend on the number of image-trained staff, the size of the department, the length of rotations away from treatment machines, and queries regarding levels of competence. A competency check should involve similar methods to the original assessment: A short period of retraining or training update, including any changes to protocol or equipment, followed by a reassessment. Each member of staff who has previously completed the imaging training and is regularly analysing on-treatment images should be involved in the regular competency check. No member of staff should be assumed to be competent at all times, regardless of grade, length of service, or working role. Staff who have previously completed the competency assessment, but who have not been involved in the analysis of on-treatment images for a period of time, should have a retraining session and a competency check before they are expected to analyse images.

The regular competency check, as well as reassessing competence in image evaluation, should address decision making following that analysis. All staff who are analysing images must be competent to implement actions following discovery of any error. An assessment of decision-making ability after discovery of an error should also form an important part of this regular competency check.

The competency check should be carried out annually or when deemed appropriate for individual members of staff. The timing may be altered due to not working with image analysis for a significant time due to rotation or time out of work. It may also be altered if a particular staff member's ability to analyse images is brought under question.

A document produced in the UK by the Society and College of Radiographers (SCoR) (2013) includes guidance on training for IGRT. Guidance includes training to be delivered in stages: Acquisition, analysis, and actions. An assessment strategy is recommended, followed by mandated regular updates to be given to all trained staff and a regular reassessment of competence (Image Guided Radiotherapy [IGRT] Clinical Support Programme in England 2012–2013).

4.3.5 Student Image Training

Imaging training for students within the radiotherapy department can cause debate. Image acquisition, analysis, and the ability to make decisions following analysis are now major and routine aspects to radiotherapy treatment delivery, and as such, newly qualified members of staff are expected to be involved in these processes. Training students in the simpler aspects of radiotherapy imaging will allow them, once working as a qualified member of staff, to become involved in imaging much quicker and reduce the training burden placed on their employer. Training in image acquisition can follow the same pattern as it does for qualified staff members; this training can be delivered alongside training in treatment delivery. As the imaging systems on most modern treatment machines are integrated with the computer systems for treatment delivery, training in image acquisition should be simple to incorporate with treatment-delivery training. Students should always be closely supervised in any interactions with patients, equipment, or documents in the radiotherapy department.

The involvement of students in image analysis must be carefully decided upon in conjunction with both academic and clinical training sites. There is a safety and quality issue with students being able to analyse real patient images on a live, working system. Safety procedures must be in place to ensure that any image review undertaken by a student is not seen as a final analysis of that image and that no corrections are made based on that registration. Image training for students may, therefore, be better placed in the academic setting; this will ensure no patient images can be reviewed by a nonqualified student. Delivering imaging training in an academic setting will have to take place on a training database. Issues associated with this not only include students not gaining a perspective of timing constraints or facing the distractions that occur when using a live system in a department, but also the issues of image gathering. Training should be delivered as close to real as possible, which will involve the use of real on-treatment images imported into a training system. Clinical departments can do this quite simply as they have access to these images without moving them outside of the health care provider's computer systems; however, transferring these images to an academic facility, often using a separate computer system, raises the issue of patient confidentiality. This must be addressed.

A common compromise is the delivery of imaging theory sessions in the academic institution to give students an understanding of the importance of on-treatment imaging, while observational training occurs whilst the student is in the clinical setting. The practical training of how to actually perform image analysis may be completed following employment as a treating member of staff. An advantage to this is that staff will have a thorough appreciation of imaging protocols, but will receive full image analysis training in their future department where they will learn equipment and protocols specific to that department.

CHAPTER 5

Concomitant Exposures and Legal Frameworks

5.1 INTRODUCTION

Since ionising radiation always has clinical consequences for a patient no matter how small the quantity, one must be careful in its use for diagnosis and treatment; always ensure there is a net benefit to the patient whenever it is used. As such, there are tight legal frameworks for ensuring ionising radiation is used wisely and that the risks of its use are suitably understood. In general terms, this means that keeping all doses As Low as Reasonably Achievable (ALARA) or Practicable (ALARP) are highly important considerations and are the basis of international guidance governing the use of radiation and protection from its use; e.g. through the International Commission on Radiological Protection (ICRP). Each country has its own legal framework within which the safe use of ionising radiation is governed, particularly for medical exposure. In the UK, this comes under the Ionising Radiations (Medical Exposure) Legislation 2017 (IRMER 2017), the legislation that governs both diagnostic and therapeutic use. Once the clinical team (in the UK, within the multidisciplinary team, or MDT) has decided to treat a patient with radiotherapy, one must be aware of all the exposures of radiation needed to plan and verify the treatment course; this can be numerous in a radiotherapy clinical pathway. In this chapter, we discuss the types of radiation exposure anticipated, which are not part of the treatment itself, the principles involved in minimising such doses, and the legal frameworks that operate to ensure these aspects that protect the patient.

For this chapter, some key references are (RCR 2008a; NRIG 2012; Ding et al. 2018; Dieterich et al. 2016a, 2016b; de Los Santos et al. 2013; IRMER 2017; and Bartling et al. 2015).

5.2 CONCOMITANT DOSE–DEFINITIONS

Concomitant exposures may be defined as "all exposures within the course of radiotherapy other than treatment exposures" (de Boer et al. 2001; RCR 2008a; IRMER 2017). This would then include all aspects of the use of ionising radiation within the radiotherapy process that are not part of the prescribed dose for therapy. It is these exposures that are deemed necessary for radiotherapy, but are not part of the dose that is delivered with curative intent to the target volumes. These exposures are still beneficial for the patient because they help plan the treatment to deliver the prescribed dose to the target volume and minimise the dose to the normal tissues and organs at risk. With on-treatment geometric verification, the benefit of the radiation dose is to gain information that helps ensure that what one plans to do with a high dose of radiation is indeed delivered as intended geometrically. Examples of these concomitant exposures would be:

- All pretreatment (planning) procedures, which would routinely be CT simulation and/or pretreatment planning scans

- **On a traditional C-arm linac:**

  - All nontreatment fields used for on-treatment imaging for the purposes of geometric verification. These could be 2-D planar images taken with the treatment MV beam; the MV beam modified for imaging purposes (usually a slightly lower energy than the treatment beam but still in the MV range); or with a separate kV source attached to the gantry of the linac

  - Volumetric imaging taken with the MV treatment beam, a modified MV treatment beam, or a separate kV imaging source attached to the gantry (e.g. MV CBCT or kV CBCT)

  - In-room on-treatment imaging sources not attached the gantry. For example, orthogonal in-room 2-D planar imaging sources (like ExacTrac) within the treatment room; or an in-room CT scanner (these technologies are discussed further in chapter 7)

- **Alternative treatment delivery systems:**

  - On advanced therapy platforms, these would include on-treatment imaging using MV fan-beam CT (using a modified MV treatment beam) to produce a 3-D volumetric image (e.g. in tomotherapy); or using an orthogonal pair of in-room 2-D kV imaging sources and panels (e.g. in Cyberknife); using MV (and, in development, kV) CBCT for volumetric imaging (e.g. in Halcyon); or using a combination of 2-D and 3-D, kV and MV imaging techniques that are mounted on an O-ring-style treatment gantry (e.g. in Vero)

On-treatment geometric verification has a different objective to diagnostic imaging and, therefore, dose considerations should ensure geometric positioning of the high dose volume for treatment rather than diagnosis. It is unlike international diagnostic programme initiatives like Image Gently (https://www.imagegently.org) or Image Wisely (www. imagewisely.org/), but still needs to balance benefit against the concomitant dose even though, compared to therapeutic doses, they are likely to be small (Dieterich et al. 2016a, 2016b). Again, within the legal framework, justification of the exposures, however small, are necessary. The overall benefit will be improved accuracy in the geometric delivery of the prescribed therapeutic dose and the planned dose distribution by better patient positioning and localisation (Jaffray and Siewerdsen 2000; Oldham et al. 2005; Thilman et al. 2006; Murphy et al. 2007). This ensures that the plan is delivered as intended, but also could enhance treatment further by permitting the use of smaller CTV-PTV margins, reducing doses to normal tissues and organs at risk still further, and possibly allowing for safe dose escalation to the target volumes (Pawlowski et al. 2010). Where internal motion or anatomical changes are considered to be likely and significant, imaging is needed in both 3-D and 4-D, with possible adaptation of the plan required through the course (Jaffray et al. 1999; Ding et al. 2008; Nijkamp et al. 2008). This type of image guidance is now the norm for modern radiotherapy in most clinical sites and for complex (but now routine) techniques like IMRT, VMAT, SRS/SRT, and SABR/SBRT (Ding et al. 2018). This places a potentially greater dose burden—and hence risk—on the patient, albeit for good clinical reasons.

The risks involved are those associated with any radiation exposure, but are mainly two-fold here for radiotherapy: The potential for secondary cancer induction and a resultant increased dose to sensitive nearby organs. Accounting for such doses into planning (Ding and Coffey 2008) and imaging protocols is important so that the magnitude of the extra concomitant dose delivered is understood and minimised whilst still gaining benefit from the on-treatment imaging procedure. Understanding the risks are important, with the ALARA principle still having merit in order to reduce the risk without significantly affecting the accuracy of target localisation. Protocols should include estimates and/or take account of the concomitant dose burden into treatment planning so that clinicians are helped in justifying and authorising all radiation exposures from a fuller knowledge of the risks and benefits involved.

Both measuring and modelling the concomitant dose from on-treatment imaging is an active area of research; reducing the dose whilst still maintaining image quality is a key aspect for the patient and service as a whole (NRIG 2012). Computer modelling of the dose from kV imaging sources has not been traditionally included within the capability of the TPS, but is under development (Ding et al. 2018) as is modelling the dose from the treatment itself to the whole body outside of the treatment field (Hauri et al. 2016). Image quality is a concept that covers both geometric accuracy and analysis/interpretation and action taken following on-treatment imaging (NRIG 2012). There is a trade-off between dose and individual image quality, so that the image is sufficient in quality to allow accurate and robust analysis and correction of set-up error. If the geometry cannot be verified on-treatment with a certain degree of accuracy, then there is a necessary increase in CTV-PTV margin required to ensure that the CTV is covered in all circumstances by the high-dose treatment volume. These are the uncertainties involved in patient positioning and the technical treatment delivery, but also the uncertainties in geometric verification on-treatment (Qi et al. 2013).

Individual image quality is highly dependent on dose (Kirby and Glendinning 2006; Morrow et al. 2012) and upon the modality and technology involved (e.g. MV vs kV x-rays; 2-D vs 3-D imaging). For the treatment course, the concomitant dose burden for the patient also depends upon the frequency of imaging, the imaging technique, the treatment technique, and the technology involved for both treatment and on-treatment imaging (Aird 2004; Waddington and McKenzie 2004; Jaffray 2005; Pouliot et al. 2005; Sykes et al. 2005; Hall 2006; Harrison et al. 2006; Morrow et al. 2012; Bartling et al. 2015; Ding et al. 2018). Imaging and dose distribution models (Hauri et al. 2016; Ding et al. 2018) can be used to assess the total burden for normal tissues and especially organs at risk close to the high-dose volume; this brings together the extra dose burden to out-of-target normal tissue structures with the total delivered dose for the course of treatment and imaging protocol (RCR 2008a; Cadieux et al. 2016; Hauri et al. 2016; Ding et al. 2018).

The risks of secondary cancer induction also need to be carefully judged; it is no longer satisfactory to assume that the cumulative imaging dose is negligible compared to the prescription dose from the treatment (Dieterich et al. 2016a, 2016b). These too need to be taken into account against the overall benefits, especially if patients are anticipated to live longer after successful treatment. Dose estimates calculated from modelling and/or deduced from measurement can be used to calculate the extra dose burden, effective doses, and, therefore, subsequent probabilities of secondary cancer induction (Lindsay et al. 2001; Hall and Wuu 2003; Aird 2004; Harrison 2004; Thomas 2004; Hall 2006; RCR 2008a; Cadieux et al. 2016). These risk estimates could become part of on-treatment imaging protocols to, for example, set action levels for frequency of imaging to limit the chance of secondary cancer induction to <0.5%, compared with a natural lifetime risk of approximately 25% (Waddington and McKenzie 2004; RCR 2008a). These could be more important for younger patients (Cadieux et al. 2016). Knowledge of individual imaging doses could help to simplify such procedures by applying a limit to the number and type of images involved, before further imaging is considered and sanctioned. In general, a threshold of 5% of the prescribed target dose could be used as a trigger for considering the dose burden in the treatment planning process; most kV-based on-treatment imaging techniques will fall within this threshold, but the principle of considering the dose burden for the imaging protocol still has merit, especially where the "heavier" imaging techniques (like daily kV CBCT or in-room kV continuous imaging) are concerned.

Individual image doses can vary quite dramatically, from a fraction of a milligray to tens or even hundreds of mGy depending upon the techniques and the technologies used (De Los Santos et al. 2013; Ding et al 2018). Most traditional C-arm technologies might use a pair of kV (or MV) 2-D images per fraction or a single kV CBCT (or MV CBCT). With technologies such as Brainlab AG's ExacTrac and Accuray's Cyberknife, the number of kV 2-D images could be considerably more (more than 80) as frequent imaging is used for monitoring patient position through relatively long treatment times (Lee et al. 2008; Ding 2015; Dieterich et al 2011; Sullivan and Ding 2015). Even for the newest of technologies, such as Halcyon, fast imaging with MV and kV energy x-rays brings a dose burden highly dependent upon frequency of use (Li et al. 2018). Each will therefore deliver an additional imaging dose to the patient's normal tissues (Murphy et al. 2007; Ding et al. 2007; Gayou et al. 2007; Wen et al. 2007; Ding et al. 2008; Ding and Coffey 2009; Ding and Munro 2013).

5.3 LEGAL FRAMEWORKS

Legal frameworks (such as IRMER 2017 in the UK) are quite clear about the justification of radiation exposure: "All medical exposures should be justified"; with the practitioner ensuring that "exposures of target volumes are individually planned, taking into account that doses to non-target volumes and tissues shall be as low as reasonably practicable (ALARP) and consistent with the intended therapeutic purpose of the exposure" (RCR 2008a; IRMER 2017).

The accompanying Medical and Dental Guidance Notes for the UK's IRR17 legislation (MDGN 2002; IRR17) state that the:

IR(ME)R practitioner responsible for the treatment exposures can justify the concomitant exposures at the outset or during the radiotherapy course, but in doing so must be aware of the likely exposures and the resulting dose so that the benefit and detriment can be assessed. This can be achieved by including likely concomitant exposures within site-specific protocols with an effective dose agreed.

In practice in most centres in the UK, the on-treatment imaging geometric verification protocols include estimates of dose for a particular site for typical verification procedures (i.e. number of images acquired through a normal course of treatment) with trigger points for further authorisation (and therefore justification) if those types/numbers exceed a certain threshold. Otherwise, the justification is included in the clinical protocol for that particular treatment, which the practitioner authorises as justified in order for treatment to commence.

The key points here are that all doses need justification, under legislation, and that those prescribed therapeutically are for eradicating the tumour cells within the carefully defined target volumes, whilst minimising and maintaining within acceptable dose constraints doses to noninvolved tissues, especially nearby organs at risk (OARs). As we recognise on a typical computerised treatment plan, optimisation of the plan tries achieve just this by taking into account (by computer modelling) exposure to the OARs from a treatment field and the high dose volume as a whole. The optimised plan also must account for in-patient scattered radiation (from irradiation of the target volumes), out-of-field leakage (e.g. through the closed leaves of the MLC), and transmission and scatter from the other parts of the linac, especially when complex, dynamic procedures are involved (such as IMRT and VMAT) for each fraction of treatment delivered. These are taken into account by the optimised treatment plan, calculated using mathematical algorithms built into the planning system that have been compared with measurements performed on the actual treatment machines within the department at the time of acceptance, at commissioning, and through the routine quality assurance programme for each machine. The computer modelling of the treatment planning system must also be commissioned (before first use and following software upgrades) so its model of the doses outside of the target volumes are precise and accurate within acceptable tolerances.

5.4 INCORPORATION INTO PRACTICAL USE

In experiments, discussion and conclusions from calculations, models, and measurements of out-of-field doses emanating from the actual treatment itself, one tries to understand and establish the full dose burden to tissues outside the target volume. They are useful for verifying the integrity of the treatment planning system in calculating nearby organ at risk doses and DVHs with the understanding that it (the TPS) can only be used with the available CT planning slices (which are rarely the extent of the whole body). They may then, in combination with the TPS, be used to assess the extra dose burden given through the use of on-treatment imaging for geometric verification, allowing total doses to be known and understood, the associated secondary cancer induction risks to be calculated, and, therefore, the overall benefit of treatment justified from the total concomitant exposure given to the patient. There must be an understanding that sometimes extra images are required (outside the normal frequency anticipated or directed within a protocol, whether it be on-line or off-line), and the doses per image should be known to allow a trigger point (in terms of concomitant dose or numbers of images) for further authorisation, if necessary, during the course of treatment. The associated risks from all concomitant doses should be regularly discussed, audited, documented within the clinical protocols or other quality system procedures, and become part of the consent process for all patients. Discussion through regular audit (RCR 2008a; NRIG 2012) with respect to national legislation and dose limits for those persons undergoing treatment is paramount. Establishing a centre's own dose reference levels that can be monitored actively through regular audit is ideal; although it is not required under IRMER 2017, it is good practice to do so (NRIG 2012).

Within the UK, the IRMER 2017 legislation has updated guidance on exposures that are "much greater than intended" (DoH 2017), bringing verification exposures into its compass (IRMER 2017, Byrne 2017). Repeat imaging, due to a failure to correctly follow a protocol or a failure in the protocol itself, now may require reporting to the appropriate legislative body if greater than intended by an appropriate multiplying factor. Examples may include an unintended verification exposure when an on-treatment image was taken when not necessary; if a daily imaging protocol was employed instead of the desired, and authorised, weekly protocol; or, similarly, if a 4-D CBCT protocol was used instead of a 3-D one. The reasoning here being that the exposure was greater than intended and could be reported dependent upon the guideline factor (Byrne 2017).

CHAPTER 6

Foundational Principles of Protocols, Tolerances, Action Levels and Corrective Strategies

6.1 INTRODUCTION

This chapter will continue on from the already outlined topic of imaging protocols to discuss correction strategies and the principles that lead to their formulation. Other aspects will be discussed, such as tolerances used and action levels, which influence the correction of imaging errors; how these affect planning and treatment margins; and why these issues need to be addressed before implementing a correction strategy. In chapter 4, imaging protocols were introduced and the contents that should be found in a typical imaging protocol were defined. This chapter will explain specific imaging protocols in more depth.

6.2 TOLERANCES

A tolerance is what will be allowed to happen; an aspect or an amount that we are comfortable with; something that does not require altering or amending:

An allowable amount of variation of a specified quantity,

Oxford Living Dictionary

In imaging terms, tolerance is the allowable discrepancy between the planned treatment and the treatment delivered. Any discrepancy that is deemed to be "within tolerance" will not be corrected. For example, if the positional tolerance for a radical lung treatment is agreed to be 3 mm following image review, any discrepancy of 3 mm or below will not be corrected; an error measured at 3.1 mm, or greater, however, will be corrected as it is outside of the agreed tolerance level.

A tolerance is required for most aspects of radiation treatment that involve patient and staff interaction. Tolerances take into account many factors affecting accuracy of treatment delivery, including immobilisation equipment used, design of treatment equipment (i.e. gantry sag, couch deflection), and organ motion (BIR 2003). Each of these factors can affect the tolerance required for radiotherapy treatments. This section will focus mainly on immobilisation and its effect on treatment tolerances. Due to the fact that all of our patients are living people, who breathe and make many different kinds of movements both voluntary and involuntary, it is not always possible to ensure every aspect of treatment is delivered precisely as planned. When working with such small margins for every patient's treatment, we can become used to considering 5 mm to be a large distance, however, if we actually stop and look at 5 mm measured on a ruler, we see that it is, in reality, a very small length. When considering all of the equipment involved in delivering radiotherapy, its size, its distance from the patient, as well as the distance a normal action like breathing can cause people and patients to move, it becomes more understandable why each treatment set-up will require a tolerance.

If any person were to be asked to keep perfectly still for a number of minutes and not move even as much as 1 mm, it would be impossible. It *is* impossible for patients to do this. We help them in this task by using immobilisation equipment. The type of immobilisation used for a patient or for a particular treatment procedure greatly influences the tolerance that can be allowed when imaging that patient. The level of stability of a patient's treatment set-up will be very different depending on the equipment used; a patient lying on a standard chest immobilisation board, with arms held back and a head rest, will be much more able to move than a patient wearing a tight fitting, individually produced mask with head and shoulder restraints (a 5-point mask) (RCR 2008a). See Figures 6.1 and 6.2.

Due to these differences in stability, different tolerance levels are required for on-treatment imaging when patients are treated using different immobilisation equipment. A tolerance of 3 mm could easily be achieved when using a customised treatment mask, but a tolerance of 5 mm may be more suitable for the patient on the chest board.

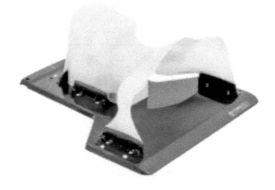

FIGURE 6.1   5-point mask.

Source: © MacroMedics BV.

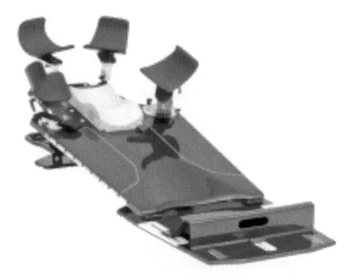

FIGURE 6.2   Chest board.

Source: © MacroMedics BV.

Examples of some immobilisation equipment and possible imaging tolerances used with each are shown in Table 6.1

When tolerances have been agreed upon for each piece of immobilisation or for different treatment techniques, it is important that every member of staff involved in either treating patients or planning patient treatments is aware of these tolerances. Treating staff must be aware of when a move in treatment position is required and when the level of error detected is within acceptable limits, therefore not requiring a correction. It is important that every treating member of staff completes the same actions and makes the same decisions regarding when an alteration in set-up is required and when the error is within tolerance. This will ensure that all patients treated in the same department are treated to the same standard and issues such as staff rotation or patients transferring between machines for treatment will not cause any changes to their treatment, positioning, or imaging actions taken. Consistency is important. Tolerances used within a department should be informed by national guidelines such as "On- target: Ensuring geometric accuracy in radiotherapy," (RCR 2008a). They can also be informed by local and national studies involving set-up accuracy of certain treatment techniques or use of individual immobilisation equipment (BIR 2003). Any staff involved in

TABLE 6.1   Possible Imaging Tolerances Associated With Immobilisation Equipment

Immobilisation device	Suggested imaging tolerance
Mattress and pillow	10 mm
3-point mask	3 mm
5-point mask	3 mm
Prone belly board	5 mm
Supine knee support and foot stocks	3 mm
SABR frame and mask	1 mm

planning radiation treatments must be fully aware of any tolerances to which the treating staff will adhere. During the planning process, staff will be involved with defining organs at risk as well as target volumes. While making decisions regarding margins around these volumes, it is vital that planning staff know what actions will be made by staff when treating these patients; for example, if the high-dose region for a target volume falls only 3 mm away from an organ at risk but the treatment tolerance if 5 mm, treatment planners need to be aware that the treating staff will not make a shift until the discrepancy is 5 mm or more. For this example, it may lead the organ at risk to receive a damaging radiation dose. In this case, planners may need to ensure there is at least a 5 mm margin before the organ at risk receives a too-high dose or inform treating staff that for this particular plan that a tolerance of 5 mm is not acceptable. Deciding tolerances for treatments is a team challenge.

The treatment aim will also have an effect on imaging tolerances. A small imaging tolerance will require stable immobilisation and will also require a comprehensive imaging schedule to ensure each treatment delivered is within that tolerance. The patient's treatment is also likely to take longer as images need to be acquired and evaluated, and any required move must be completed before treatment. This process is not always suitable, especially for those in pain or uncomfortable in their treatment position. Patients undergoing palliative care, quite often, are not able to lie in the treatment position for long times; consideration of this must be made when deciding margins and imaging tolerance to be used for these patients. Palliative treatments usually involve lower doses of radiation, as the aim is not to eradicate the tumour, sometimes making treatment fields less conformal and margins often larger. For these reasons, the imaging tolerance may be increased depending on the anatomical site of the treatment and surrounding organs; many radiotherapy departments use an imaging tolerance of 1 cm where there is no dose consideration required for surrounding tissues. This will ensure that strict immobilisation is not necessary, the patient can be kept more comfortable, and the likelihood of a lengthy treatment process is minimised as treatments are less likely to require a correction after image analysis.

Tolerances must be used as we cannot guarantee treatment accuracy to be exact. All staff, treating and planning, must be aware of tolerances and actions that should be taken following image analysis. Immobilisation must be considered when deciding tolerance levels for treatments. The last and maybe most important consideration is patient ability and cooperation. Is the patient physically able to maintain a position, and is it necessary for them to do so?

6.3 MARGINS

Margins used for radiotherapy treatment were introduced in chapter 1. This section will discuss how these margins and imaging tolerances have an effect upon each other.

GTV: Gross Tumour Volume, includes visible or measurable tumour extent

CTV: Clinical Target Volume, surrounds the GTV to give a margin for microscopic, immeasurable spread

PTV: Planning Target Volume, surrounds CTV allowing for potential errors in treatment placement. (ICRU 1993)

The size of the expansion from the CTV to produce the PTV will be greatly influenced by a number of factors including technical uncertainties (such as, gantry sag and treatment couch deflection), dosimetric uncertainties of the treatment beam, and the reliability of the treatment set-up (ICRU 1993, Stroom et al 2002). The reliability of the treatment set-up is the point at which on-treatment imaging can be used to reduce the CTV to PTV margin if a reliable imaging protocol and procedure is in place (Skarsgard et al 2010, Moreau et al 2017). A smaller PTV margin around the CTV will reduce the amount of healthy tissue in the treatment volume and, therefore, reduce side effects for the patient. Very conformal margins can only be successfully used for treatment with reliable imaging processes and small imaging tolerances. If the errors due to treatment placement cannot be minimised through on-treatment imaging, then larger planning margins must be used to ensure complete CTV coverage.

Conventionally, any margins used will be equal around the extent of the target volumes, a 5 mm CTV to PTV margin will mean the full CTV is extended out, in all aspects, by 5 mm, as shown in Figure 6.3.

However, if a section of that PTV encroaches in to an organ at risk (OAR), the margin may need to be altered, see Figure 6.4.

Altering the treatment planning margins in such a way will affect the imaging tolerance for that plan. Directional guidance can be given by the planning staff to the staff members responsible for treating and imaging. For the example in Figure 6.4, such advice may be a standard imaging tolerance of 5 mm for all directions, however, in the lateral plane, RT/LT, a 2 mm tolerance in the left direction should be applied. This will ensure that any discrepancies in the left direction will be corrected when an error greater than 2 mm is detected, not 5 mm, as all other directions.

Some specialised treatments may use a very conformal PTV margin. Treatments such as stereotactic radiosurgery utilise much higher doses per fraction and so must keep the exposure to healthy surrounding tissues to a minimum. The margins for these treatments are very conformal to the GTV with a minimal PTV margin used (Kirkpatrick et al 2014,

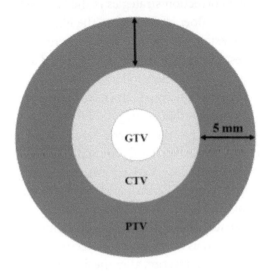

FIGURE 6.3   Showing an equal 5 mm margin in PTV expansion.

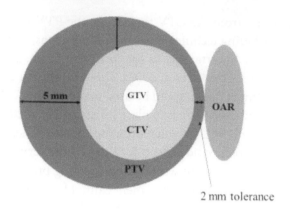

FIGURE 6.4   Showing a nonuniform PTV expansion.

Kocher et al 2014, Naoi et al 2014). Imaging is vital for such techniques. There is likely no imaging tolerance used for these treatments, meaning that any discrepancy between planned position and treatment position will be corrected. Specialist immobilisation equipment will be required for such treatments. Images are likely to be acquired before treatment commences and at numerous points throughout treatment delivery to ensure there has not been any intrafraction movement, i.e. the patient has not moved during treatment delivery.

6.4 CORRECTIVE STRATEGIES

When imaging equipment and its intended use within a department have been established, along with tolerances and margins to be applied to images and treatments, correction strategies are to be implemented.

6.4.1 NAL Correction Strategy

There have been many correction strategies discussed and investigated for radiotherapy. One of the most widely used correction strategies is the "No Action level" (NAL) strategy. This was first put forward by De Boer et al in 2001. The NAL strategy involves repetition of imaging for three fractions to establish whether the error is random or systematic. Following results from the three images, an average amount is calculated and that shift is applied to all subsequent treatments. Images are repeated, following the initial treatment shift for two further treatments to ensure any initial discrepancies have been corrected by the shift applied.

Example of using the NAL correction strategy, see Table 6.2.

For the NAL correction strategy, the average error discrepancies will be corrected. In Table 6.2, the average errors calculated will convert into the isocentre shifts applied to the patient's treatment set-up for all subsequent treatments, i.e. for patient 1 the isocentre will be moved 2.1 mm sup, 0.3 mm Rt and 1.5 mm anterior; this should correct the errors seen on the first three days of imaging. The patient will then have subsequent images taken, usually for the next two treatments, to ensure the correction has been implemented correctly. The NAL correction strategy was developed as a way to ensure accuracy of patient position as early in the treatment process as possible. Correction after the first treatment alone was considered to be too early in the treatment process as there would be no possibility of determining whether

TABLE 6.2   Example of NAL Correction Strategy

	Patient 1 (3 mm imaging tolerance)		Patient 2 (3 mm imaging tolerance)	
Treatment 1	2.5 inf	0.7 Lt	3.8 sup	4.5 Rt
	1.5 post		0.3 post	
Treatment 2	2.0 inf	0.1 Rt	4.1 sup	3.6 Rt
	2.0 post		0.1 Ant	
Treatment 3	1.7 inf	0.4 Lt	3.5 sup	4.2 Rt
	1.3 post		0.3 ant	
Average error	2.1 inf	0.3 Lt	3.8 sup	4.1 Rt
	1.5 post		0.0 ant	

(all measurements in mm)

the error was random or systematic. To make a precise, mathematical decision regarding these isocentre moves for a patient's treatment, the more points of information included, the more accurate the results will become (Bel et al 1993), i.e. a decision whether to make a shift for a patient's treatment will be more accurate if image analysis results for 20 treatments are included, rather than just one or two. However, this would mean that the patient has received 20 treatments with no correction applied to any errors noticed. There has to be a compromise between the desire to include as many imaging results as possible and the need to correct any errors as early as possible in the patient's treatment delivery (Bortfeld et al 2002). The recommendation is to include three images (De Boer et al 2005, RCR 2008a); all patient's will have the first three days' images analysed, and the average of these three results will be used to inform the moves to be applied for all subsequent treatments.

6.4.2 eNAL Correction Strategy

Following on from the NAL correction strategy came the eNAL strategy (De Boer et al 2007). The original NAL correction investigated and corrected for any errors occurring with the set-up and delivery of the initial radiotherapy treatments. It did not account for any further errors that may be seen, variations in patient weight or size, tumour progression or shrinkage, internal organ motion, or other factors that can lead to a patient's treatment position altering throughout the treatment course (De Boer et al 2001). For this reason, the extended NAL (eNAL) strategy was developed. The initial verification of the patient's position is evaluated using the same NAL process, and the extension applies later in the treatment course as the image process is repeated. The usual interval between imaging is one week, so in standard treatment schedules this would mean images taken every five treatments. Should the image analysis completed after five treatments show no errors outside of agreed tolerances, treatment continues with no further alterations and no further imaging for another five treatments.

Example of eNAL strategy, see Table 6.3

Following the eNAL correction strategy, patients were imaged at treatment number 10. For patient 1, all of the evaluation results are within the tolerance, therefore no further actions are required. This patient will have further images taken following another five treatments, i.e. treatment number 15. Patient 2 however, was found to have a treatment error following

TABLE 6.3 Example of eNAL Correction

	Patient 1 (3 mm imaging tolerance)		Patient 2 (3 mm imaging tolerance)	
Treatment 1	2.5 inf	0.7 Lt	3.8 sup	4.5 Rt
	1.5 post		0.3 post	
Treatment 2	2.0 inf	0.1 Rt	4.1 sup	3.6 Rt
	2.0 post		0.1 Ant	
Treatment 3	1.7 inf	0.4 Lt	3.5 sup	4.2 Rt
	1.3 post		0.3 ant	
Average error	2.1 inf	0.3 Lt	3.8 sup	4.1 Rt
	1.5 post		0.0 ant	
	All average shifts applied for subsequent treatments			
Treatment 4	1.2 inf	0.2 Rt	0.5 sup	0.4 Lt
	0.1 ant		0.1 ant	
Treatment 5	0.8 inf	0.1 Lt	0.2 sup	0.7 Lt
	0.2 post		0.2 post	
	eNAL protocol followed			
Treatment 10	0.7 inf	0.2 Lt	0.4 sup	0.7 Lt
	0.2 post		3.7 post	
Treatment 11	No further imaging required		0.2 sup	0.3 Lt
			4.2 post	
Treatment 12			0.1 sup	0.7 Lt
			3.9 post	
Average error, #10, 11 and 12	Within tolerance		0.2 sup	0.6 Lt
			3.9 post	

(all measurements in mm)

image analysis at treatment 10. This patient then required further imaging to establish if the error was random or systematic and the degree of correction required. Images were taken for three treatments, an average result acquired, and further shifts applied to the patient's isocentre. This patient will have further images acquired at treatments 13 and 14 to ensure corrections have rectified the error. Following this, the imaging process will be repeated at treatment 19, i.e. five treatments after the last imaging process is completed. The eNAL strategy is the weekly repeating of the NAL process to ensure no changes to the patient's position throughout the whole treatment process. It is hoped that by using the extended imaging strategy, the patient's position can be analysed and corrected if needed throughout the course of radiotherapy instead of during only the initial few treatments.

6.4.3 Absolute Correction and Adaptive Radiotherapy (ART)

As treatment plans and margins have become more conformal to the target area, there have become more treatments where the NAL and eNAL correction strategies are no longer applicable. If the treatment involves very small margins, such as a brain treatment; contains mobile structures, such as a prostate treatment; or organs likely to alter in size or shape, such as a lung treatment, not imaging for five treatments may not be acceptable. Following the eNAL

TABLE 6.4   Example of Daily Imaging and Full Correction, ART

	Patient 1		Patient 2	
Treatment 1	2.5 inf	0.7 Lt	3.8 sup	4.5 Rt
	1.5 post		0.3 post	
	All errors fully corrected			
Treatment 2	2.0 inf	0.1 Rt	4.1 sup	3.6 Rt
	2.0 post		0.1 Ant	
	All errors fully corrected			
Treatment 3	1.7 inf	0.4 Lt	3.5 sup	4.2 Rt
	1.3 post		0.3 ant	
	All errors fully corrected			

(all measurements in mm)

protocol, there will be no scope to evaluate the treatment position for the four treatments that occur in between the imaging schedule. This may result in compromising treatment position and treatment outcome during these sessions. A more recent correction strategy being used is daily imaging and full correction of any errors detected before treatment delivery. Adaptive radiotherapy (ART) is the process of monitoring treatment variations, either due to position of the treatment or geometry of the target volume (Yan et al 1997). ART involves daily imaging of the treatment position and the daily correction of any errors seen, replanning of the treatment may also occur if the target volume changes, meaning the treatment plan is no longer optimal (Schwartz et al 2011, Brouwer et al 2015).

For the following discussion, focus will be toward the daily correction of treatment position. Using ART, all random errors will be discovered prior to treatment delivery as every treatment will be verified. Following NAL or eNAL, random errors in treatment placement occurring on nonimaging days will not be seen.

Example of daily imaging and full correction, see Table 6.4

This method would appear to be the simplest of correction strategies as there is very little decision making needed; all errors, regardless of amount, are corrected following image analysis. There is no discussion regarding when to image a patient, as an image will be taken prior to every treatment being delivered. The dose received from the daily imaging must be considered. This schedule of imaging is most suited to the imaging methods requiring the lowest radiation dose; the benefit to the patient, in terms of accuracy and smaller margins available, must be considered alongside the increased dose received whilst acquiring these daily images.

6.5 ACTION LEVELS

An action level for any measurement is the point at which an action is required (RCR 2008a). Action levels can vary in what action is expected and different action levels can be used for the same patient or treatment set-up. Action levels are different from tolerances. The first action level will be outside of the agreed tolerance but smaller than the gross action level. For the first action level, treatment may be allowed to continue with a repeat of imaging before the next fraction. For the gross action level, no treatment should be delivered until

TABLE 6.5   Examples of Action Levels

Site	Tolerance	Action level	Action	Gross action level	Action
Radical chest	5 mm	7 mm	Reimage before next treatment	1 cm	Do not treat, check set-up parameters, correct error before treatment delivery
Radical prostate	3 mm	5 mm	Reimage before next treatment	7 mm	Do not treat, check set-up parameters, correct error before treatment delivery
Palliative chest	8 mm	1 cm	Do not treat, check set-up parameters, correct error before treatment delivery	1 cm	Do not treat, check set-up parameters, correct error before treatment delivery

any error of this magnitude has been corrected. Correction may be checking basic set-up instructions to ensure all moves have been made correctly or checking identification issues such as the correct patient and the corresponding treatment details have been used. It may not always be appropriate to simply correct the positioning of the treatment by applying isocentre shifts for an error at this level.

Example of appropriate action levels, see Table 6.5:

As discussed previously, tolerances for each treatment site are decided based on a number of factors, including planning margins, immobilisation, and treatment intention. Action levels are set similarly. The gross action level is the level set in order to identify and correct a gross error before any treatment delivery. The CTV to PTV margin does not allow for any treatment delivered above the agreed gross action level, therefore, any error of this magnitude may result in an underdose to the CTV and overdose to organs at risk adjacent to the target volumes. Any treatment delivered within the gross action level will still ensure the CTV has received adequate treatment and surrounding tissues have not exceeded dose constraints. See Figure 6.5.

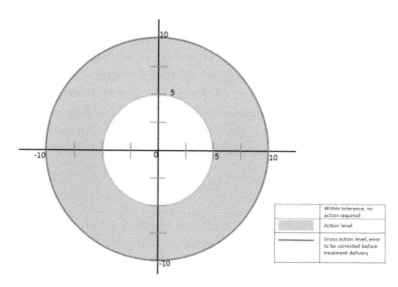

FIGURE 6.5   Relationship between no action required, action level, and gross action level discrepancies.

Figure 6.5 shows a treatment target with a 5 mm surrounding tolerance, any analysis of images with results of 5 mm or less will not require any further action. Imaging results of between 5 mm and up to 1 cm, will require action, as the result is less than the gross action level. The action required may not be immediate, however, and could be investigated further at the next treatment. Any error discovered to be 1 cm or more falls outside of the gross action level and therefore immediate action must be taken—no treatment is to be delivered before this error is corrected. If using a correction strategy that involves correcting all errors regardless of magnitude, action levels will not apply as there is no agreed tolerance for that treatment, i.e. no positional discrepancies will be tolerated. The cause of any gross errors discovered during on-treatment imaging, however, should be investigated.

II

Technology and Techniques in Clinical Practice

CHAPTER 7

Imaging Science, Imaging Equipment (Pretreatment and On-treatment)

7.1 INTRODUCTION

This chapter continues the exploration of imaging science concepts, first introduced in chapter 2, since they are fundamental concepts for any type of medical imaging—especially those associated with ionising radiation and the use of x-rays. We then examine the main pretreatment imaging technologies, with a view to understanding their basic use in providing reference information for both target volume and organs at risk (OAR) delineation for planning, and for the production of reference images as the basis for on-treatment geometric verification imaging. We conclude with a detailed description of the on-treatment imaging technologies used with traditional C-arm linacs, including those based on nonionising radiation technologies.

7.2 IMAGING SCIENCE CONCEPTS

Here we delve a little deeper into some of these concepts for the sole purpose that they help us understand the basic features of all images used in radiotherapy. By doing so, we gain insight into what we can and, perhaps, cannot "see" with images and understand the advantages and limitations of current imaging used for pretreatment simulation and planning (which is the main reference point for on-treatment imaging and geometric verification). We begin here by expanding upon basic image science concepts, and then concentrating on those particular to x-ray imaging. For greater depth, the reader can find some of the fundamental concepts covered extremely well in papers examining the imaging science behind electronic portal imaging devices (EPIDs), including work by Boyer et al. (1992) and Munro and Kirby (2004).

7.2.1 Spatial Resolution

The characteristic of spatial resolution helps us understand how well small objects can be observed within an image; it is an index of the detail that can be perceived, the ability to see small objects and edges, effectively the sharpness of an image (see Figure 7.1). Quantitatively, we can measure spatial resolution by observing how closely lines of different widths can be to one another and be resolved by an observer in the final image. For x-ray imaging, these are formed by materials that are widely different in their attenuation properties. In pretreatment imaging, good spatial resolution is important for observing boundaries between tissues and organs, for example, outlining target volumes or organs at risk. An image with high spatial resolution has the potential to enable the observer to view the boundaries between tissues more easily since the edges are more clearly visible.

These tasks may be partially or fully automated now by automatic image processing and analysis software and even through machine learning; their success (in terms of how accurately the detailed tissues can be outlined automatically) depends upon the mathematical algorithms and methods used and the quality of the imaging process, which includes the spatial resolution of the image.

For on-treatment imaging, this is important in image matching and registration (for the algorithms or as observers ourselves) to be able to distinguish edges to compare and match the on-treatment image (in 2-D or 3-D) to the reference image with confidence and calculate the set-up error associated with it. This might be related to bony features or to soft-tissue anatomy.

Where fiducial markers are concerned, poor spatial resolution may mean we cannot observe the fiducial markers with accuracy or even at all; so the uncertainty in performing the match (and hence the uncertainty in measuring our set-up error) is high. The consequence may be whether we perform an appropriate or inappropriate correction in for example an on-line procedure. This may mean the correction made to set-up is larger or smaller than that required in reality. The same difficulties are encountered by software designed to automatically detect the fiducial markers (in both on-treatment and pretreatment images)

FIGURE 7.1   A schematic diagram showing the result of an image detector with good, high spatial resolution (Image 1) and one with poorer, low spatial resolution (Image 2). The result is that the detail can't be resolved in Image 2—the detail in a radiograph will look blurred or fuzzy.

and make the appropriate correction. For off line procedures, it may mean an inappropriate calculation of any systematic error involved.

7.2.2 Contrast

The *Oxford English Dictionary* defines contrast as, "The state of being strikingly different from something else in juxtaposition or close association." This is quite a general definition that could apply in many different spheres—contrast in an image, or contrast in tastes, emotions, hair colour, etc. With a medical image, we are concerned with the observer's ability to resolve one anatomical object from another when they are next to or near one another; there is a better ability to distinguish the clinical object from the background or other anatomical features if the object is in contrast with them.

If there is no visual contrast, we are not able to resolve that object since we can't distinguish it from the background or other objects. For x-ray imaging (whether planar 2-D x-rays or 3-D CT procedures), this only happens if there is a difference in attenuation coefficients between an object and its background. If the attenuation coefficients are too similar, even though anatomically and clinically there are two quite separate objects within the patient, there is little image contrast between them (because they both attenuate the x-rays by similar amounts), which makes it difficult to resolve the two in the final image (see Figure 7.2).

This is a problem for distinguishing the boundaries of organs at risk in, say, the CT scan images used for pretreatment simulation and planning, or for distinguishing features on on-treatment images when performing image-matching with the reference image. Within treatment planning, this may result in the OAR receiving more dose than necessary or underestimating the volume of the target and, therefore, under-dosing it. For on-treatment geometric verification, the result might be an appropriate or inappropriate set-up correction, either giving more dose than planned to an OAR or not adequately covering the target volume with the high dose volume.

FIGURE 7.2 A schematic diagram to show the result of differing contrasts. At left, the high contrast and the object (the smaller square) can be seen readily with respect to the background (the larger square). At right, there is very low contrast—one can barely make out the smaller square with respect to the larger one.

FIGURE 7.3   An example of noise within the image. The pattern is random and conveys no useful information about the object (say an anatomical object). In reality, this variation in texture is not actually part of the object one is trying to image—it is caused by other sources.

7.2.3 Noise

Noise may be regarded as all the random fluctuations that obscure or do not contain meaningful information within an image. Often, more than half the differences seen in an image come from noise in the image data. Noise is considered a major limiting factor that inhibits image quality in CT imaging as it is the proportion of the object observed within an image that contains no true information about that object. Noise is characterised visually by a grainy appearance within the image or other structural details (see Figure 7.3). If we were able to cut open the patient and observe the anatomical organs with our own eyes, that graininess is not real—noise draws a picture of something that isn't there in reality.

7.2.4 Signal-to-Noise (SNR) Ratio

Here is an index that compares the level of a desired signal (within an image that would be the anatomical detail we wish to see) to that of the background level of noise (the data in the image we don't wish to see since it has no true information about our object) (see Figure 7.4). SNR and noise are important clinically because they affect how confidently we can observe parts of an image (either to diagnose accurately and then make an accurate treatment plan, or to geometrically verify with confidence the positioning of the patient for delivering that plan). If the diagnosis is inaccurate, we may miss the opportunity to treat a patient of their disease, or treat the wrong disease altogether.

If, within simulation and treatment planning, we cannot identify features with confidence for our target volumes, more often than not a decision is made to treat a larger volume (to be more certain of irradiating the malignant cells). This could lead to giving more dose to nearby OARs and/or having to compromise on the prescribed dose to limit toxicity, which might then reduce the probability of local tumour control.

FIGURE 7.4   A schematic to show how the visibility of an object depends upon the signal-to-noise ratio. At left, the signal (the anatomical object) is much stronger than the noise component making the object is fairly easily visible. At right, however, the noise is as strong as the signal, making the object very difficult to see in the image.

Better, higher quality imaging (through better spatial resolution, higher contrast, reduced noise and as high a signal-to-noise ratio as possible), leads to greater confidence in outlining what needs to be treated and, therefore, a potentially smaller margin. A smaller margin can potentially reduce toxicity and allow for a safer escalation of dose and, therefore, offer a greater probability of local tumour control. It also leads to more accurate assessment of set-up error in the on-treatment setting and greater confidence in delivering the treatment plan as intended. Better quality imaging also helps in reducing concomitant dose through the reduction of repeat imaging.

7.3 CHARACTERISTICS OF NOTE FOR X-RAY IMAGING

Some of the key characteristics specific to x-ray imaging relating to the quality of images are:

- X-ray attenuation
- Scatter
- 2-D projections
- Subject contrast and x-ray energy
- Signal-to-noise ratio
- Spatial resolution
- Detective quantum efficiency

7.3.1 X-ray Attenuation

As noted above, in x-ray imaging, contrast is dependent upon how two anatomical objects attenuate the x-rays passing through them. X-rays are attenuated by each object, meaning energy is removed from the beam by the x-ray interactions that take place, which are mostly the photoelectric effect, Compton scatter, and pair production. The number of photons (intensity or fluence) that pass through decreases in an approximately exponential fashion following the familiar equation, see Equation 7.1:

$$N = N_0 e^{-ux}$$

Eqn 7.1

Here, x is the thickness of the object (in mm or cm) and u is the linear attenuation coefficient (in respective units of mm^{-1} or cm^{-1}). The linear attenuation coefficient depends upon the composition of the object AND the energy of the x-rays. Even if the two objects are, anatomically, different (different tissues or organs), if they have similar attenuation coefficients, the x-rays passing through each will be attenuated in the same way; the number striking the image detector will be similar and, therefore, there will be no contrast between the two on the final image. The observer (or even a computer algorithm) will not be able to distinguish between the two.

7.3.2 Scatter

Scatter is radiation that arises from interactions of the primary radiation beam with the atoms within the object being imaged. Compton scatter, where an incoming photon interacts with an atomic electron as if it is essentially free, thus producing a high-energy electron and a scattered photon (of lower energy than the incoming photon), is the primary contributor to scatter in x-ray imaging. Because scattered radiation deviates from the straight-line path between the x-ray focus and the image detector, scattered radiation is a major source of image degradation in x-ray imaging no matter the modality. In diagnostic imaging (planar x-rays and conventional CT), this can be reduced by the use of scatter grids, which are thin grids made of attenuating material (lead or steel) with "septa" focused back to the radiation source.

For on-treatment imaging with the MV treatment beam, one cannot use scatter grids because the energy of the x-ray is too high. Nor are they usually used in kV CBCT applications, so scatter is a source of degradation in image quality (even with kV energy x-rays) in the on-treatment imaging setting; the image quality of kV CBCT images is generally poorer than that of conventional planning (fan beam) CT images because of this added scatter component.

7.3.3 2-D Projection Images

Planar or 2-D imaging is the most traditional imaging technique, both diagnostically and therapeutically. But we are limited in the amount of detail or information one can obtain on the image because it is only one single projection. The image that results is a shadow pattern of all the objects that attenuate the x-ray beam as it passes through. If there is little attenuation, then more x-rays pass through and strike the detector; if there is a large amount of attenuation, then fewer x-rays pass through and we get contrast in the image between the different objects, which gives us an image. However, the attenuation that occurs is from all the different objects that the x-rays pass through, such as different anatomical objects of different densities, different compositions, and different attenuation coefficients. The resulting attenuation is an average of all those objects. The information, therefore, from the different objects is compressed and merged into the planar 2-D image. This image still produces valuable anatomical information (plain 2-D x-ray images are still widely used in diagnostic imaging), but it is limited compared to a volumetric (3-D) method like CT. In addition, contrast is reduced because of the presence of scatter from the patient.

7.3.4 Subject Contrast and X-ray Energy

One of the most crucial factors to consider when comparing images taken with diagnostic energy x-rays (usually tens of kV) and therapeutic energy x-rays (6 to 20 MV) is that of subject contrast. This is the difference in contrast of the same anatomical object observed at different energies (beam qualities) of x-rays. The same anatomical object or objects will "look" different depending upon the energy of the x-rays used to acquire the image. This is because the attenuation coefficients are dependent upon both the composition of the tissue AND the energy of the x-rays used. So, even though the object is the same (it is the same patient, after all), the contrast between it and the background will be different; in some circumstances so different that it is no longer distinguishable from the background. Consequently an image (of precisely the same part of the same patient) will **always** have a different quality when acquired with kV or MV energy x-rays because the physics is simply different—no matter what type of image detector is subsequently used, the subject contrast will always be different.

Looking a little deeper conceptually, let's consider an ideal situation where x-rays pass through a homogenous material within which is an anatomic object that has a different attenuation coefficient to the homogeneous material. If the average number of x-ray quanta detected as an image background is B, and the number detected behind the anatomic object is O, then it can be shown (Boyer et al. 1992) that the subject contrast S is given by Equation 7.2

$$S = 2\big[(B-O)/(B+O)\big] \qquad \text{Eqn 7.2}$$

Further derivation shows that the subject contrast naturally depends upon the difference in attenuation coefficients between the object and the anatomic background and also the fraction of x-ray quanta scattered by the material. If the difference in attenuation coefficients increases, there is greater subject contrast; if the fraction of x-ray quanta scattered decreases, then again, there is greater subject contrast

As a simple example, consider a 1 cm thick section of cortical bone surrounded by 20 cm depth of water. Simple calculations will show us that the contrast (between the bone and the surrounding water in the image) changes with energy since the attenuation coefficient of the cortical bone is dependent upon the energy of the x-rays. We find that at:

- 50 kV: The subject contrast is approximately 18.5%

- 1 MV: The subject contrast is approximately 1.5%

- 6 MV: The subject contrast is approximately 1.0%

This is because at the lower kV energies, the predominant x-ray interaction is the photo-electric effect, which has a probability of occurrence highly dependent upon atomic number Z (it is proportional to Z^3). So there is much greater attenuation of the x-rays within the 1 cm slice of bone, since the difference in Z between the bone and the surrounding medium (water) is relatively large. The approximate effective atomic numbers of different materials are shown below:

- Air: Z = 7.78

- Muscle: Z = 7.64

- Water: Z = 7.42

- Bone: Z = 12.31

Since the atomic numbers are so different between bone and water (or muscle or air), there are many more photoelectric interactions (since it is proportional to Z^3) that take place in the bone compared to the surrounding water. The attenuation coefficients are very different and, therefore, there is a large subject contrast observed (nearly 20%). This is why diagnostic radiographs, acquired using kV x-rays, show bony features extremely well.

Now consider changing the energy to the megavoltage range—say about 6 MV. Now the predominant interaction is Compton scatter, an interaction that is independent of Z and dependent primarily on the electron density of the material. Now, a large difference in attenuation of the x-rays between the bone and the surrounding water (and hence a large subject contrast) will only be observed if the densities between the bone and the surrounding material are different. Approximate densities of these materials are:

- Air density = 1.205 Kgm^{-3}

- Muscle density = 1040 Kgm^{-3}

- Water density = 1000 Kgm^{-3}

- Bone density = 1650 Kgm^{-3}

Now we see that the difference in density is not so great between the bone and the surrounding water (or muscle). Therefore, the difference in attenuation coefficient isn't nearly as great and the contrast is much, much less (it is poorer). So bony detail is not so easy to "see" in an image taken with 6 MV x-rays compared with one taken with kV x-rays. In fact, it is easier to see air compared to muscle or bone (there is a much greater contrast), because the density of the air is approximately 1/1000 that of surrounding muscle or bone. So it is easier to see, for example, bowel gas within the pelvis or the air in the lungs surrounding a lung tumour with MV energy x-rays than with kV energy x-rays.

The key take-home message is that subject contrast is always different between images (of the same object, the same part of the patient) when an image is taken with kV energy x-rays compared to one taken with MV energy x-rays. This is crucially important when considering the different on-treatment imaging options, and comparing those taken with the MV treatment beam and those taken with a separate kV x-ray source. There will always be a difference in image quality—the kV energy image being better in terms of contrast than the MV—no matter how good the imaging detector might be.

7.3.5 Signal-to-Noise Ratio (SNR)

This is one of the most important parameters for x-ray imaging and we have examined it in general terms above. For clinical, x-ray-based images, it can be formerly defined as the "ratio of the signal difference between an anatomical structure and the background, to the statistical noise associated with the detection of the x-ray quanta necessary to form the signal." We are examining, in clinical terms, the ability to resolve an anatomical feature (more often than not in on-treatment imaging, these are the edges, points, curves and detail of bony features, soft-tissue structures or even fiducial markers) from within the noise or graininess we might observe in the image. Some of this graininess is due to the quantised nature of the x-rays, some will be random noise from all the various conversion processes (from the initial detection of the x-ray photon) involved in forming the electrical signal in the pixel of the display monitor, and still some will be structural features associated with the flat-panel imager itself and the ability of the system software to calibrate this out. The noise, in essence, constitutes all those aspects of the image that are not really part of the anatomy we are trying to observe; the higher the SNR, the better the quality of the image and the more easily the observer can identify appropriate anatomical structures and boundaries in on-treatment imaging.

Again, if we wish to consider it a little deeper, mathematically, using the same terms as in section 7.3.4 (letters B and O), it can be shown (Boyer et al. 1992) that the signal-to-noise ratio (SNR) can be expressed by Equation 7.3:

$$SNR = (B - O) / \sqrt{(B + O)} \qquad \text{Eqn 7.3}$$

In imaging science terms, it is dependent upon:

- **The size of the object:** For clinical imaging, we are limited by the anatomy of the patient, which cannot easily be changed.

- **The subject contrast:** As defined above, this is greatly dependent upon the x-ray energy being used. The move to on-treatment imaging using kV energy x-rays (from a separate source to that used for treatment delivery), has enabled a better exploitation of subject contrast.

- **Photon (x-ray) fluence:** We are limited by dose to the patient. The limitations are greater when using the MV treatment beam, which generally delivers a higher imaging dose than with the use of kV x-rays. But for more continuous imaging, intrafractionally, even the use of kV x-rays can be a significant dose burden for the patient.

- **Efficiency of the detector:** We are limited by technology. Later in this chapter, we shall see how improvements in technology can improve image quality through improved SNR (among other factors), by the fact that the imaging detectors can make better use of the available signal, i.e. the x-rays striking the panel after passing through the patient.

7.3.6 Spatial Resolution

Two of the main parameters that influence spatial resolution (the fineness of the detail perceived in an image) are x-ray source size and the detector elements size within the imaging detector as a whole (primarily the pixel size).

- If the x-ray source size is large, the penumbra at the edges of objects becomes larger and the edges of anatomical detail become more blurred. The source size for kV x-ray tubes is generally smaller than that used for the MV treatment beam, so spatial resolution is often better for kV-based on-treatment imaging.

- The detector size (the size of the pixels) can be a limiting factor in terms of spatial resolution. The smaller the pixel size, the higher the spatial resolution. In the era of radiographic film, the spatial resolution was extremely good, since the "pixels" were the size of the chemical crystals (or grains) involved (often 0.01 mm or less in size). For flat-panel imager detectors, the digital pixel size is usually between 0.1 and 0.5 mm. Generally, the smaller the pixel size, the better the spatial resolution, although other factors need to be taken into account to make a direct relationship to better image quality on the whole; if the pixels are too large, then the observer struggles to resolve and identify small objects or the fine detail of anatomical edges.

7.3.7 Detective Quantum Efficiency

The detective quantum efficiency (DQE) of a detector gives an overall measure of the signal-to-noise transfer characteristics of the imaging detector as a function of spatial frequency. In essence, the parameter describes how efficiently the whole imaging system (when considering first detection of the x-rays passing through the patient all the way through to the formation of the image on a monitor) conveys information, particularly signal-to-noise ratio, through to the final image observed. Highly efficient systems (with a high DQE) will do this well (without significant loss of detail) at all spatial frequencies, i.e. when observing either very fine detail in the anatomical image or broader features across the whole image.

7.4 PRETREATMENT IMAGING EQUIPMENT

Pretreatment imaging equipment (equipment used to either diagnose the condition of the patient or for treatment planning purposes once a decision has been made to use radiotherapy) is mainly used for informing the clinician (usually from a range of imaging modalities and tests) about the target volume (GTV and/or CTV) that needs to be treated.

It is needed to give a good, clear, and accurate picture of all the anatomical structures that might contain tumour cells and, therefore, need to be within the high dose volume to receive a tumouricidal dose of radiation. Each patient will have a variety of imaging examinations and scans; some specifically for planning purposes, but all for gaining as much anatomical and functional information about the patient and their condition as possible. In this way, the treatment plan can be tailored to the individual so that the maximum possible dose is delivered where needed, with as little as possible to the normal surrounding tissues and especially those organs nearby known to be particularly sensitive to radiation (nearby OARs).

7.4.1 Radioisotope Imaging

This is a key imaging modality in diagnosis, staging, and planning of radiotherapy, particularly of note, too, for treatment follow-up. A crucial difference from the more straightforward x-ray-based imaging is that radioisotope imaging is **functional**. Rather than just being able to demonstrate the internal anatomy and only observe anatomical features (as with basic 2-D plain x-ray, CT, CBCT, basic MR etc.), radioisotope imaging allows one to gain important clinical insight into the function of organs and tissues. The best diagnosis comes from multimodality imaging, a combination of functional and anatomical imaging, so the clinician is better informed about the position and extent of disease and set-up to choose their target volumes with better confidence with respect to the malignancy. From such information, planned target volumes could increase or decrease compared with those based purely on anatomical imaging, or indeed the treatment intent might change (from localised and radical to more systemic and/or palliative).

In terms of on-treatment imaging for geometric verification, radioisotope images are not used directly as reference images. However, they may have been used to inform the drawing of target volumes on CT images, which are then used as reference for 2-D and 3-D on-treatment x-ray imaging.

7.4.2 Positron Emission Tomography

Positron emission tomography (PET) is another form of functional imaging used to inform, in particular, of the potential of malignant cells in the body. It relies on the use of radioisotopes, which are positron emitters (rather than mainly gamma emitters as used for standard radioisotope imaging), and uses subtly different technology inside the PET scanner as compared with a gamma camera. When a positron emitting isotope (such as Fluorine 18) is chemically bound to a glucose analogue (e.g. as in fluorodeoxyglucose, FDG), it becomes a means for targeting cells and tissues that have a high metabolic function and process glucose and other sugars to get the energy for proliferation. This is the case for all cells, but malignant cells do so at a much more rapid rate than normal tissues and differentially use up more glucose than normal cells. By detecting the increased radiolabeled glucose metabolism with a high degree of sensitivity, PET potentially can identify cancerous cells, even at an early stage, when other imaging modalities may miss them. PET imaging, therefore, can be used to inform better the size and positioning of the planning target volume for radiotherapy.

The inherent difficulty with both radioisotope and PET imaging is that they suffer from a relatively poor spatial resolution. So they require the use of good anatomical imaging (such as CT and MR) to inform geometrically where within the patient's anatomy there might be increased uptake and, therefore, the tissue volumes that may need to be within the high dose volume. On separate PET and CT scans, for example, this can be done "by eye" by the trained clinician in a qualitative manner, but without the submillimetre accuracy required for modern radiotherapy techniques. Registering electronically separate sets of PET and CT (or MR) images is also difficult because of the poor spatial resolution of the radioisotope images. Registration is possible, but only with quite an associated degree of uncertainty, resulting in larger target volumes. The ideal way around this is a technological combination of scanners, i.e. the design and use of, for example, a PET-CT scanner, as we will see in the following sections.

As with traditional radioisotope imaging, PET images are not used themselves as a reference image for on-treatment imaging. But they may have been used to inform the drawing of target volumes on CT images (or MR images), which are then used as reference for 2-D and 3-D on-treatment x-ray imaging. Moreover, with advent of PET-CT (and PET-MR) scanners, the functional information is registered directly into the anatomical images, which may then be used as reference for on-treatment geometric verification purposes. This is provided the PET-CT scanner has been equipped with a flat couch top and can accommodate scanning the patient within the immobilisation equipment that will be used for the course of treatment.

7.4.3 Computed Tomography—3-D

X-ray computed tomography is the bedrock of pretreatment imaging and computerised treatment planning, and the main basis of the reference images on which on-treatment imaging currently relies. Current generation CT scanners use a kV x-ray source mounted diametrically opposite a series of detectors placed in multiple slices on a slip-ring gantry. These rotate around the patient at speeds of up to four to five rotations per second, whilst the patient on the couch moves smoothly and continuously through the bore of the machine, thus making the acquisition of image data helical. Typically, energies (on single energy units) are approximately 80 to 140 kV; modern computer processors allow the systems to reconstruct images as they are being acquired. The acquired data from the multiple rows of detectors enables the attenuation coefficients of the tissues and organs within the body to be computed and converted into CT numbers (or Hounsfield units) that are mapped onto a scale range so that water has a CT number of 0 and air a CT number of –1000. Because radiotherapy is delivered with MV energy x-rays (which deposit energy in the human body primarily through Compton scatter interactions), the CT numbers are mapped theoretically to electron density for computerised treatment planning purposes; this mapping is confirmed stoichiometrically by scanning a phantom with inserts that have very well-defined and known electron densities.

The resulting images have a high spatial resolution (usually a fraction of a millimetre) and very good contrast resolution and SNR. However, for different organs and tissues to be differentiated from one another (especially at tissue boundaries), they must possess different attenuation coefficients as we have discussed above. For this reason, soft tissues (and the boundaries between them) are often difficult to resolve. But the images are geometrically correct (a perfect geometric representation of the anatomical structures in the body) and thus ideal for treatment planning; they are excellent for outlining target volume structures and normal tissues, especially those that are more radiosensitive and regarded as OARs in the planning process.

We use CT as the main source of reference images for on-treatment verification imaging. Once the treatment plan has been completed (with identified positions for the treatment isocentre and beams), reference images can be exported from the treatment planning system (or virtual simulation software) in the form of 2-D or 3-D datasets, depending upon the on-treatment imaging requirement. Two-dimensional images are in the form of digitally reconstructed radiographs (DRRs) or digitally composited radiographs (DCRs). DRRs are constructed using the computed attenuation coefficients from the CT scans for the exact same geometry (scaling and orientation) as a real 2-D x-ray radiograph that could be taken of the patient. It can be computed for any treatment orientation combination of gantry and couch positions, so it mimics the set-up used on the C-arm linac for on-treatment imaging (with either kV or MV x-rays).

For DRRs, the scan width and spacing of the individual CT slices should be as fine as possible to produce a good, acceptable spatial resolution in the cranio-caudal direction and avoid "stair-step" artefacts on anatomical structures on the DRRs. They must also be exported with information as to the treatment field edges and/or isocentre position in order to compute set-up errors on treatment. Optionally, they may also include the target volume and OAR outlines for comparative purposes.

For 3-D on-treatment imaging, the reference dataset is the full 3-D set of CT scans exported from the treatment planning system or virtual simulation software, again with information as to the isocentre position for the checked and approved treatment plan. They are often exported with the target volume and OAR outlines for comparison during geometric verification in 3-D or in 4-D. This information is crucial for adaptive radiotherapy (in either 3-D or 4-D).

7.4.4 Computed Tomography—4-D

One of the largest problems for 3-D computed tomography is that of motion, when taking images of the body where there is likely to be internal anatomical motion—most notably the thorax—whilst the patient is free-breathing. Movement artefacts in 3-D CT imaging occur because the patient is driven continuously on the couch unidirectionally through the bore of the scanner in order to acquire a 3-D dataset from individually acquired slices. During this process, if there is significant internal motion in the cranio-caudal direction, this results in banding artefacts where the tumour volume and other moving structures (like the diaphragm and lungs themselves) are not imaged completely, resulting in incomplete identification of the tumour volume. The result is that it is difficult to delineate the target volume with confidence and parts of the tumour volume could be outside the high dose volume unless large margins are used. This is a greater problem when the tumour is located in parts of the lung that have the greatest amplitude of motion due to breathing, for example, the lower lobes rather than those in the most superior aspects of the lung or those nearest to more rigid central lung structures.

One way of reducing this problem is to take 4-D CT scans whilst the patient is free-breathing. For this the breathing phases of the patient (from maximum inspiration to maximum expiration) are identified during the acquisition of the scan slices; the acquired projections and/or slices are then 'binned' into the associated breathing phases before reconstruction (using maximum intensity projection) into appropriate planes for the different phases of the breathing cycle. The results can be played back in a movie sequences to show the motion and identify the actual movement of the target volume (especially the position at maximum/minimum breathing phases) so that the target volume, and therefore the high dose volume for treatment, can be identified more accurately.

To do this, one must use a breathing "surrogate" to identify the breathing phases. This could be via an external device placed on the patient's chest that is monitored optically using IR light and linked electronically to the position of the couch as the CT scans are being acquired. Another option is a pressure belt wrapped around the patient's midriff to monitor the diaphragm position. Alternatively, a physical surrogate is not necessary if actual movement of the internal anatomy itself can be identified accurately and linked to couch position (Zhou et al. 2018). In all cases, the acquired slices are linked to particular phases of the breathing cycle.

From the binned slices, an internal target volume (ITV) can be identified from the various delineations of the GTV/CTV on the slices for each maximum/minimum identified positon of the tumour volume in the cranio-caudal direction; this way the full, accurate dimensions of the tumour volume present, and an accurate map of its position during respiration, can be identified and included in the high-dose planning volume. Taking an average of the CT slices from the different breathing phases reconstructs a set of CT scans that are equivalent to what is observed during 3-D free-breathing but without any missing positional information; it is a true 3-D average of the moving structures. This set can be used for dose calculations in the TPS to model free-breathing treatment delivery.

For the purposes of 2-D on-treatment imaging for geometric verification, the outputs needed are the appropriate DRRs computed from the average 3-D scan set and the delineated volumes, such as ITV and PTV, for overlay purposes for identifying the expected extent of tumour motion. These are computed for the appropriate gantry angle to be used for the 2-D on-treatment images (whether MV or kV). For 2-D methods, it must be appreciated that these will be "snapshot" images during a particular part of the breathing cycle.

For the purposes of 3-D on-treatment imaging, the outputs needed are the average 3-D CT scan set and the delineated volumes, such as the ITV and PTV. These become the reference images for comparison against 3-D on-treatment imaging (e.g. kV or MV CBCT), with the outlines again giving good information as to expected tumour motion. When CBCT is used as the 3-D on-treatment imaging modality, the volumetric data is acquired over a number of breathing cycles (since one full rotation of the gantry is limited, by legislation, to 1 minute as opposed to approximately 0.25 seconds for the rotation of the CT gantry); so the CBCT shows a true average of the breathing motion.

Four-dimensional on-treatment volumetric x-ray imaging, 4-D CBCT, is also now possible (Hugo and Ruso 2012; Zhou et al. 2018); raw data projections are sorted into the appropriate breathing phase and linked to a breathing surrogate or internal anatomy. Doses can be lowered by limiting the number of projections, but this necessarily reduces image quality. Motion compensating algorithms (Rit et al. 2011) can also be used to make the CBCT images appear clearer, although this can sometimes be with a loss of fine anatomical detail, but with a good removal of streaking artifacts associated with internal anatomical motion (Hugo and Ruso 2012; Rit et al. 2011; Zhou et al. 2018). Combined with 4-D CT gating or active coordination methods, the treatment itself can also be gated, with the beam interrupted when outside a predefined tolerance window. The required reference images from the TPS could also include the CT scan sets from the individual breathing phases (or at the least, those for maximum inspiration and minimum expiration) for comparison with the on-treatment acquired 4-D volumetric data. However, a clinical difficulty here with free-breathing lung techniques is that the high dose volume is quite large; it is required to accommodate the range of movement of the target volume. Ideally, one would like to reduce this volume since a smaller volume would irradiate less of the normal lung tissue surrounding the tumour and reduce toxicity for the patient. As a result, the prescription dose could be escalated, thereby affording a greater chance of local tumour control and overall survival for the patient.

Some ways around this are to limit the magnitude of the motion by using Deep Inspiration Breadth Hold (DIBH) or Active Breathing Control (ABC) techniques so that scanning and consequent delineation of the target volume only takes place when there is less movement or indeed no movement of the tumour volume; treatment is delivered only when the tumour volume is in the same position as planned during deep inspiration or when the breath is held, thus enabling a reduction of the overall target volume. This reduction of the target volume, and therefore the irradiated volume, can be achieved with free-breathing by gating the treatment delivery (i.e. the beam is only initiated when the tumour volume is in a particular part of the breathing cycle) or by tracking the tumour volume during its motion with technologies such as Cyberknife (see chapter 10).

7.4.5 PET-CT Imaging

As we have seen, PET scans can give us excellent functional information of the tissues and organs within the body (most notably, the possibility of the presence of tumour cells), but the imaging modality has poor spatial resolution (for most scanners, this is 3 to 4 mm). CT scans can give us excellent anatomical information with very high spatial resolution (usually less than 0.5 mm). The ideal planning method is to combine these two modalities to allow the clinician to use both functional and anatomical information together to decide and then delineate the target volumes for treatment. When the images from these two modalities are presented as two independent sets, acquired as separate imaging sessions on separate pieces of equipment, there is the inherent difficulty of registering and overlaying one set of data on the other (the functional PET data onto the anatomical CT data) with any real confidence of positioning, particularly when the spatial resolution of the PET data is much poorer than the CT. Tissue and organ boundaries are very poorly defined on the PET scans and therefore the registration cannot be performed with great confidence, meaning margins of error are still large, on the order of millimetres at least. The ideal is a more confident procedure or technology for doing this, which is why PET-CT scanners have been designed and are now in widespread use.

PET-CT scanners combine in one technological unit, both a high quality CT scanner and a PET scanner, with a common couch passing through the centre bore of each. So for a single visit and single imaging session, a patient can have high-quality functional and anatomical imaging scans without being repositioned between each scan; the patient remains in the same set-up position on the couch for each scan. The two scanners share a common frame of reference room coordinates and therefore registration of the two data-sets is straightforward and submillimetre; the margin of error in accurately combining the two sets of scans is reduced considerably. PET scan times are generally reduced from about 45 to 60 minutes down to approximately 25 minutes because there is excellent anatomical detail available from the CT scan acquired. This means that motion artefacts (from a pro-longed procedure) are reduced. The PET scan data is often converted into colour-wash and overlaid automatically on the CT scan data. The fused images can be imported into the TPS for target volume outlining and delineation. For the purpose of on-treatment imag-ing for geometric verification, the outputs are the same as above for CT and PET; they are primarily DRRs and CT scans from the CT data (for 2-D and 3-D on-treatment imaging respectively) and accurate volume outlines from the combined data sets.

7.4.6 Magnetic Resonance (MR) Imaging

MR imaging is unique in the field of large 3-D (and 4-D) imaging datasets in that it does not use ionising radiation. By the use of very high-strength magnetic fields, radiofrequency waves, and the inherent spin properties of atomic nuclei, MR imaging can produce very high quality anatomical images with considerably better soft-tissue image quality than, for example, CT imaging. As such, soft-tissue boundaries between organs and tissues are more readily identified, reducing the uncertainty of delineation for both target tissues and OARs. It brings forward the potential to reduce planning margins (which would naturally be larger, if there is uncertainty of the involved tissues) and, therefore, use more conformal dose distributions and even begin to dose escalate whilst maintaining or even reducing normal tissue toxicity.

However, there are complications and challenges with using MR images alone for treat-ment planning and then as reference datasets for on-treatment imaging. The images are not necessarily geometrically sound, i.e. their geometry (size, scaling, aspect ratio) are not always equivalent to reality (unlike CT images). There are elements of distortion in the images, often dependent upon the size of the anatomical subject within the bore of the scanner and with respect to the main, fixed magnetic field. This is a problem for treatment planning (which is crucially dependent upon accurate depths, areas, and volumes of tissues to model treatment dose distributions) and for producing appropriate reference imaging for geometric verification on-treatment, which must be geometrically free from distor-tion for accurate set-up correction. Much work has been conducted in understanding and correcting for such distortion with the result of many commercially available distortion correction algorithms being available, bringing spatial integrity to within 1 mm of reality for many clinical sites.

A second complication is that, since the images are acquired without ionising radiation, there is no inherent data that can be readily converted into electron density for dose computation purposes. This is primarily a challenge for treatment planning. Bulk correction methods (delineating organs and tissues and assigning a most-likely electron density to those volumes) are available and constantly improving in their viability. If electron density is readily available on near-distortion-free images, one could envisage an era of MR only simulation, planning, and guidance for a patient workflow—a subject we will examine in chapter 10. Methods are also being developed whereby CT equivalent images (simulated CT slices and DRRs) can be generated from MR simulation and planning images for the prime purpose of on-treatment geometric verification methods, since these are mostly x-ray based using kV and MV x-rays in 2-D, 3-D, and 4-D methods.

However, MR imaging is already in considerable use in treatment planning, by means of fusion with CT images (in a similar way to fusing and registering PET and CT images). Both modalities have very good spatial resolution (MR approximately 0.5 mm to 1 mm), excellent contrast resolution, and very good signal-to-noise ratios (although this is dependent upon the acquisition time for many MR imaging sequences). Automatic fusion of CT and MR datasets is now possible for many clinical sites, with little need for manual intervention, although the veracity of the fusion is always checked by treatment planners before proceeding. Outlining by the clinician of target volumes and by planners of organs at risk is then achieved by delineation on the MR images, with automatic reproduction onto the CT images. This allows the higher quality MR images to inform delineation better for soft tissues, especially the boundaries between them (for example between prostate walls and bladder and/or rectum for prostate treatments). Treatment planning, dose computation and optimisation, generation of DRRs, etc. can all then take place in the known CT space as normal. This allows MR to inform in the planning process, but the reference images for on-treatment imaging for geometric verification are DRRs and CT slices as per a CT only workflow.

7.5 ON-TREATMENT IMAGING ON TRADITIONAL C-ARM LINACS

7.5.1 Why the Need for On-treatment Verification Imaging?

The rationale behind on-treatment imaging and geometric verification has already been covered in the earlier chapters. The main purpose is to "close the loop." Having used a wide variety of imaging and other clinical information to decide how best to treat the patient with radiotherapy and delineate the planning target volume (the volume of tissue that we are going to irradiate to a high dose so that we can guarantee as much as possible that we cover the clinical target volume with a tumouricidal dose of radiation) and organs at risk (OARs), on-treatment verification imaging then helps assure us that the treatment will be delivered as planned, geometrically delivering the treatment fields in the correct place, within an acceptable tolerance, for every fraction. Imaging at the time of treatment delivery has been at the heart of this for decades. Historical reviews of the development of portal imaging (as it was first called) and on-treatment imaging can found in research by Langmack (2001), Antonuk (2002), Herman (2005), Kirby and Glendinning (2006), IAEA (2007), Evans (2008), and de Los Santos et al.(2013).

Within the literature, one finds many different terminologies for what is broadly described now as on-treatment imaging for geometric verification: Image guided radiotherapy. The original term was "portal imaging," since it was an image acquired on-treatment using the treatment port, the treatment field. This was initially a rectangular field, with perhaps some shielding using lead blocks or a shaped field designed using a low-melting point alloy like Cerrobend. It was not until the late 1980s that linacs with an integrated Multileaf collimator (MLC) became available; and not until the turn of this century that such linacs started to become the norm across all departments in the UK with associated portal (MV) imaging equipment. Now, all modern linacs sold come with an MLC and usually both kV and MV imaging technology.

As the technology changed, the terms MV imaging and electronic portal imaging were defined, and the modern device for acquiring an on-treatment image became the electronic portal imaging device, EPID, a term still used today in various ways. The addition of a secondary source of x-rays (a diagnostic kV energy x-ray source) was trialed in the 1980s (Biggs et al. 1985), but was only made viable when the image detector technology changed and the commercial companies designed equipment with it incorporated onto the C-arm linac; what we know as on-board imaging (OBI) or x-ray volume imaging (XVI) and similar terms now. Volumetric imaging became possible with the development of the algorithms and the detectors to make cone beam CT (CBCT) a reality. With the move to other forms of imaging device within the treatment room (including nonionising radiation methods), perhaps the best current terms are electronic patient imaging or on-treatment imaging to describe all the imaging that can take place within the treatment room, either immediately prior to the delivery of the treatment fraction, between individual treatment fields, continuously during the treatment fraction during beam on, or at the end of the fraction.

7.5.2 Evolution of On-treatment Imaging

The origins of on-treatment imaging and their evolution into what has become known as image guided radiotherapy, as the process for using on-treatment imaging as the principal guidance mechanism for the delivery of radiotherapy, are well captured in the articles listed previously; the original premise was that once a patient had been set-up on the treatment couch, images were taken using the MV treatment beam and treatment portal (hence the original name of portal imaging). The rationale for doing so, but now with more sophisticated 2-D, 3-D and even 4-D methods, is still the same today. Two chief methods were used then: (1) using the first few monitor units of the treatment, analysing the image and then making any corrections to set-up before continuing with that particular treatment field called a localisation image; and (2) acquiring an image as in (1) or throughout the entire treatment field, analysing the image, and then making any necessary set-up corrections on the following treatment fraction or after the first few fractions, which is called a verification image. Analysis in both cases was of the position of anatomical features (usually bony features, as the only ones that were barely visible in the MV image) with respect to the field edges or the isocentre. The field placement error was the geometric difference between what was observed (anatomically) at the time of treatment and the simulation reference image. Again, the literature serves us well in describing such early developments in detail, specifically work by Herman (2005), Kirby and Glendinning (2006), Evans (2008), van Herk (2007), RCR (2008a), de los Santos et al. (2013), and Zou et al. (2018). The primary concern was to remove any systematic error (see chapter 1) from the treatment set-up, since this set-up error has the greatest clinical impact on both target volume and OAR doses (Herman 2005; van Herk 2007; RCR 2008a).

These would equate approximately to what we now call on-line and off-line imaging, as described in earlier chapters. However because of the detection media available (film), these methods were slow and limited; both for the on-treatment image and the pretreatment simulation image (acquired on a simulator using a diagnostic kV x-ray source). Patients were mostly planned entirely on a treatment simulator under either fluoroscopy and/or using 2-D planar kV images. Reference images were obtained for on-treatment verification later, or, if available, planned using a CT scanner, but often with a limited number of CT slices (because of operational restrictions in the exposure from the x-ray tube; usually resulting in heat overload if too long an exposure was used). Computerised reference images (e.g. DRRs) were of very poor quality, mainly because of the limited number of CT slices that could be acquired for a pretreatment planning scan and at spacings of usually 10 mm or even greater. So after computerised planning, patients were taken to the simulator to verify the planned treatment and obtain a kV image (usually on film) that could be used as reference for on-treatment (portal) imaging and geometric verification.

The main purpose of on-treatment imaging was to analyse set-up error of the treatment position with respect to the planned position of the patient for each field and ensure that the patient was being treated, geometrically, as planned. Other advantages were also noted, such as verifying the size and shape of the treatment field. Before Record and Verify and the Oncology Management System were established as they are now, setting up each field was a manual task or a manual selection of the appropriate field, often on both the linac and a separate MLC computer. Without appropriate checks, it was quite possible to select different patients on each, and still initiate radiation. Portal imaging allowed, mostly, interfractional movement to be analysed—set-up errors from one fraction to the other—to then calculate systematic errors. The image also provided a permanent record of the actual treatment itself.

The earliest methods used radiographic film. It is useful, historically, to understand how our current methods have evolved (so one might encourage similar thought processes for the future development of current electronic methods). An image was acquired with the film placed naturally in a light-tight cassette. After exposure, the film required minutes of processing before the image could be viewed, usually with the film placed on a light box. Often the film cassette was mounted on a separate stand (rather than being on the gantry, always opposite the treatment beam). In addition, efficiency of detection was poor, so although film has a very high spatial resolution (better than our current electronic methods), the subject contrast was limited (by MV subject contrast anyway), and the display contrast was fixed by the development process. The dynamic range (shades of dark to light) was fixed, so if it was over or under exposed, the only recourse was to repeat the image the following day; no electronic image enhancement was possible and more than anything else it was slow. There was no possibility of having an instant image, so off-line methods were those most used; any on-line work required the patient to be on the treatment couch for minutes between acquiring the image, processing it, analysing it and then making any corrections to the treatment couch for proceeding with the treatment field—and that was just for one field.

The earliest design changes to improve the detection efficiency of the film are worth noting because they are presently at the heart of our current technologies for both MV and kV imaging with a flat-panel detector. Changes were made to the film cassette for portal imaging (and also for simulator pretreatment imaging) by introducing a metal plate on the front surface and front/back phosphors so that the principal mechanism within the detector was firstly in producing high-energy electrons (through photoelectric and Compton interactions) by the MV energy x-rays interacting with the metal plate. Then the high-energy electrons interacted with the phosphor to produce optical photons (visible light) that then exposed the film. This mechanism is still present in a similar form in our flat-panel imager technology used today.

The first step in moving toward devices that would produce a more instant image was the first generation EPIDs. Two main types were developed— a liquid ion chamber array, and a camera-based fluoroscopic EPID. For the camera-based EPID, the initial interactions were similar to what has just been described; x-rays exiting the patient interacted with a metal plate (steel or copper) to produce high-energy electrons, which then interacted with a phosphor to produce visible light. This light was then bounced off a front-silvered mirror into a video camera—initially analog, but later digital CCD type cameras—very similar to the ones in our modern phones. The camera would then relay or digitise the image and it was displayed on a monitor, captured and stored for further analysis and recording of the on-treatment geometric verification.

The first major advantage here was that the image was instant, available within a few seconds of exposure, making on-line analysis and set-up correction a possibility. Image quality was as good as film (although the quantum efficiency wasn't necessarily the same), mainly because the images could be digitally manipulated and processed; brightness and contrast (window level and width) could be adjusted to help correct over- or under-exposure; images could be acquired with a few monitor units (five to 10, or less), with both localisation and verification (short and long) exposures possible. Fields were acquired either solely using the treatment portal, or in combination with a wider open field (double exposures) or using a wide-open field on its own (Kirby and Glendinning 2006; RCR 2008a). It also brought to play other imaging methods including real-time (or movie loop) imaging and digital manipulation and combination of images (e.g. composite imaging) (Kirby et al. 1995). Set-up error was analysed in the same way as film (with respect to the field edges or a central axis marker) and on-line/off-line protocols were designed.

7.5.3 Initial Challenges

Numerous limitations and challenges existed with the initial film methods and even the earliest EPIDs. But the EPIDs brought in an era of real, instant imaging with the versatility of digital processing. Image comparison and registration was possible within computer software interfaces for comparing on-treatment images with digital simulator images or the TPS generated DRR, which also had improved in quality. The earliest studies showing the clinical efficacy of using on-treatment (portal) imaging (effectively demonstrating the clinical need for it) were achieved with some landmark patterns of care studies using film and the earliest EPID devices.

Efficiency of detection within the EPID was the main problem for the early generation EPIDs, especially the camera-based systems. The relatively considerable distance between the visible light emission from the scintillator and the camera meant that the conversion efficiency was poor and introduced other noise sources, too (Kirby 1996), thus lowering the signal-to-noise ratio and the overall image quality. It was also limited by subject contrast since the image was still acquired only using the MV treatment beam, even though research was active in modifying the treatment beam to make it more "kV-like" by lowering its MV energy and trying to introduce low-Z target materials. The treatment beam was still the x-ray beam of choice, since images acquired were of the true treatment isocentre—that of the MV beam.

7.5.4 The Active Matrix Flat-Panel Imager (AMFPI)

A solution came in the form of the active matrix flat-panel imager, which is now the basis of our current x-ray based imaging methods on C-arm linacs and most other treatment technologies. Antonuk (2002), Herman (2005), Kirby and Glendinning (2006), and Evans (2008) excellently described the technology and its initial use—most notably comparing the improvements in imaging with the MV beam with earlier generation EPIDs. The improved contrast, reduced noise, and higher detection efficiency all result in a better SNR ratio and better image quality for on-treatment imaging.

The devices developed (as they are today) were indirect detection devices, very similar to that of the first generation EPIDs and, indeed, to that of film within a cassette. Incoming x-rays (usually the treatment beam or a larger field, but using MV x-rays) exiting the patient would firstly interact with a metal plate to produce high-energy electrons, which then interacted with the phosphor to produce optical light photons. The main difference for the active matrix flat-panel imager (AMFPI) occurs now at this point—where the optical photons interact immediately (there is no large geometric path length to consider here) with the photodiode (amorphous silicon) layer through the photoelectric effect. This produces photoelectrons, which then discharge the capacitors in each pixel (see Figure 7.5). The charge in each pixel is proportional to the intensity of x-rays falling upon it. Electrical signals are read sequentially from each pixel, row by row, and fed into a computer for acquisition, display, and storage. The image can be processed, manipulated, and analysed (compared with a pretreatment reference image) for computing set-up error in an identical manner to the first generation EPIDs. Indeed, manufacturers changing hardware from first-generation EPID to AMFPI technology produced identical or very similar software interfaces for the user.

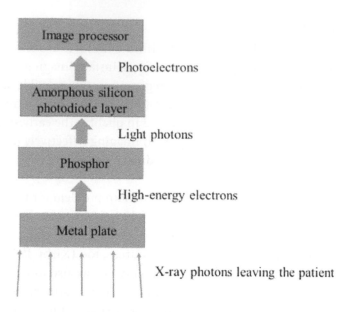

FIGURE 7.5   A schematic of the fundamental layers within an AMFPI and the typical interaction processes that happen within them. Incident x-ray photons are converted into high-energy electrons, then to optical photons, and finally into photoelectrons that affect the charge stored in each of the pixels, thus forming an image.

The AMFPI was therefore a panel detector, similar in physical attributes to typical film cassettes, but producing high-quality electronic images quickly. It is still used now for both kV and MV imaging (with some minor modifications) on the linac. A typical detector array has the following characteristics:

- 2-D array of photodiodes, usually a minimum of 1024 × 1024 detectors (pixels)

- A real-time digital camera, able to acquire image frames at a minimum of 3 frames per second

- Dynamic range is a minimum of 16 bits (over 65,000 shades of grey)

- The sensitive imaging area is about 40 × 40 cm, with the pixel spacing approximately 0.4 mm apart

Clinical uses were very similar (if not identical) to those of earlier EPIDs with the use of planar, 2-D MV imaging. The AMFPIs could be used with both kV x-rays for diagnostic imaging and pretreatment imaging (e.g. on a treatment simulator), and also for on-treatment imaging, initially with just the MV energy x-ray treatment beam. Set-up errors could be detected by registering, matching, and comparing on-treatment and pretreatment reference images (either a simulator image or a DRR). Image acquisition methods were very similar (short exposures, long exposures, double exposures, composite, and real-time imaging) using similar or fewer MUs (1 to 3 MU). Transit or exit dosimetry also became more viable with the improved signal-to-noise ratio and linear response of the AMFPI, although the foundational work on this area of in vivo dosimetry was established with the earlier generation EPIDs (Kirby and Glendinning 2006; van Elmpt et al. 2008).

The addition of a separate, kV energy x-ray source and its own imaging panel (mounted, for two out of the three main commercial C-arm linacs, at 90 degrees to the main treatment arm of the linac) made 2-D imaging possible with better quality and greater similarity to pretreatment images (since both used kV x-rays). Volumetric (3-D) kV cone-beam CT became possible with the development of appropriate computer algorithms for reconstructing a 3-D image dataset from the individual 2-D kV projections at each gantry angle using a wide-area detector (the AMFPI) and a cone beam of x-rays (rather than a fan beam as used in conventional diagnostic and planning CT scanners).

7.5.5 Current Methods With MV and kV X-rays—2-D Imaging

7.5.5.1 Surrogates

When a patient's treatment is planned, the clinician will decide on the anatomy to be encompassed with the high dose volume, which normal structures are at risk (especially those more sensitive to radiation), and the dose tolerances allowed for these. This might be done from a general view of the anatomy (from 2-D DRRs or, more likely now, from CT image slices following CT scanning and/or simulation). Depending upon the clinical case, this would likely be a GTV and/or CTV from which margins are then applied to produce a PTV to be covered with the prescription dose from the beam arrangement. The most crucial geometrical point here is the placing of the treatment isocentre(s) with respect to the target volumes.

For on-treatment geometric verification, imaging is performed to ensure that, anatomically, the structures outlined are treated as planned, which means that one needs to verify that the relationship of the clinical structures to the isocentre is the same between planning and treatment. However, it is rare that precisely the same imaging modalities, energies, and technologies can be used for on-treatment purposes as for pretreatment simulation (as we have seen in chapter 2). So, depending upon what anatomical structures are visible at planning and for the type of on-treatment imaging available, a choice of structure is needed for verification and an assumption made that clinical structures observable during on-treatment imaging are in a known geometric relationship to the target volume structures outlined by the clinician at treatment planning. This is where the concept of surrogates is required—if, on treatment, one cannot identify or "see" the same structures as one can when outlining the target volumes, an alternative is required.

A classic point is that 2-D on-treatment imaging will **never** produce the same level of anatomical detail as 3-D imaging at the time of planning; so compromises are needed and other structures must sometimes be used.

7.5.5.2 Surrogates—Skin Markers (Alone) and Bony Anatomy

The most traditional and original surrogate, which is still used in a number of cases, especially in the palliative setting, are marks on the patient's skin. These are tattoos marked during simulation (now usually from CT simulation) indicating the position of the treatment isocentre, or reference marks from which the treatment isocentre can be established by 3-D reference moves of the couch. Just prior to treatment delivery, the patient is set-up with respect to the machine isocentre and further images may or may not be taken. When no images are acquired on treatment (which is the case for very few clinical sites now), the assumption is made that the relationship between the target volume and the skin marks is identical to that established at the time of CT simulation and planning. This is not always the case, even for quite straight-forward treatment geometry.

In a similar way, bony landmarks may be the surrogate used. In this instance, skin reference marks are still used with or without reference moves, but on-treatment images are acquired and compared against reference images (usually DRRs for 2-D imaging and CT slices for 3-D imaging) (see Figure 7.6). As we have seen in the imaging science sections, subject contrast is highly dependent upon the energy of the x-rays being used for on-treatment imaging. So, for most 2-D techniques, bony features are far more visible than soft tissue structures. Although the soft–tissues are usually the target tissues for most clinical sites, geometric verification can only be achieved by using anatomical objects that are visible for both reference and on-treatment imaging. Bony features are the structures used as the surrogate for the tissues that are, in reality, the target volume. There is, therefore, the assumption that the target volume soft tissues and the bony features are in the same geometric relationship at the time of CT simulation and planning and then at treatment

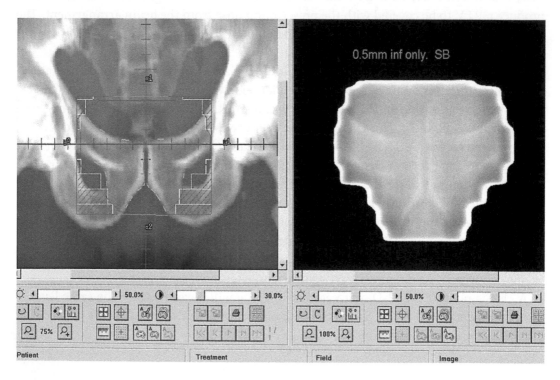

FIGURE 7.6 The computer interface from Elekta's on-treatment verification imaging system for 2-D MV EPID imaging and geometirc verification using bony anatomy mathcing. On the left is the DRR with bony features outlined; on the right is the MV image for this pelvis patient.

Image courtesy of Elekta.

delivery. Again, studies have shown that this is not necessarily the case; internal organ motion can cause a significant difference (for some clinical sites in tens of millimetres) between planned and treatment positions when bony features are used as the surrogate.

For consistency of approach and more accurate geometric verification, stable anatomical points are required for identification on the reference DRRs or CT slices and then on the on-treatment 2-D (kV or MV) images or 3-D (kV or MV) CBCT images. National guidance, such as the "On-target" publication (RCR 2008a), indicates recommended anatomy for a number of clinical sites.

7.5.5.3 Surrogates—Implanted Fiducial Markers

A fiducial marker is defined as something that could be used as a standard of reference or measurement—an object placed in the field of view of an imaging system that appears in the image produced, to act as a point of reference or measurement. For some clinical sites, it is possible to improve our geometric knowledge of the soft-tissue target structures by implanting fiducial markers within them since we cannot observe the soft-tissue structures themselves and/or they move differentially with respect to nearby bony structures. These markers are made of substances (usually metal) with a high contrast, with respect to the background tissues, for the energies of imaging involved. Usually stainless steel produces acceptable contrast for kV energy x-rays; gold is required for suitable contrast for imaging with MV energy x-rays (Mutanga et al. 2012). Recent research has proposed fiducials in a liquid/gel form (de Blanck et al 2018) that can be used in a number of clinical sites with suitably high-contrast imaging properties. The surrogates represent the position of the target volume structures within which they are implanted. Usually two or three markers are used so that acquiring images from any angle gives a unique image of their positioning. Assumptions are made that once implanted, they (a) do not migrate significantly with respect to the target tissue and (b) do not migrate significantly with respect to each other. Thus, any movement of the target tissue is reflected by movement of all the fiducial markers acting like a rigid body. They are used mainly for prostate treatments (see Figure 7.7) and also for some lung tumours. In a similar approach, surgical clips, placed postresection following, for example, lumpectomy in breast cancer patients, can be used for increasing visibility for partial breast irradiation of the surgical bed. Steel clips are easily visible for kV energy x-ray imaging.

7.5.5.4 Image Reference Points—Measuring Set-Up Error

In all cases outlined above, the surrogates used are delineated or identified on pretreatment imaging modalities, such as CT slices from CT simulation or the planning scan. During treatment planning, target volumes are outlined and the isocentre positioned appropriately for the plan, usually in the centre of the target volume. The treatment beam angles are chosen and the plan optimised, either through forward or inverse planning, depending upon the treatment technique. For VMAT, the number of arcs is chosen, so individual beam directions are not needed. In all cases, the target volume structures (and the OARs) have a known geometric relationship with respect to the treatment isocentre. The structures to be used as surrogates (bony anatomy or fiducial markers) are therefore also in a known geometric relationship with respect to the isocentre of the plan. To deliver the plan

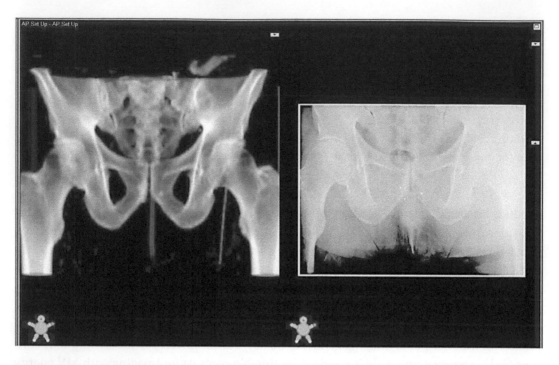

FIGURE 7.7 The computer interface from Varian's on-treatment verification imaging system for 2-D kV imaging with implanted fiducial markers. On the left is the reference DRR from the planning CT scans; on the right is the kV image for this pelvis patient, with the three gold see markers clearly visible.

Image courtesy of Varian Medical Systems.

as intended, the isocentre of the treatment machine must be placed in the patient's anatomy as planned allowing the surrogates to accurately localise the target anatomy on the treatment machine. Reference images are exported from the TPS (or VSim software) for use on-treatment. For 2-D on-treatment imaging, these are DRRs reconstructed for the appropriate gantry angles at which on-treatment images will be acquired. They will show the surrogates—the reference anatomy or the positions of the fiducial markers visible (or likely visible) on-treatment with respect to the isocentre—ready for comparison with on-treatment imaging.

When on-treatment images are acquired in 2-D (whether with the MV treatment beam or a separate kV imaging source attached to the gantry), the reference anatomy or the fiducial markers are identified within comparison software on the linear accelerator or within the oncology management system (see chapter 13). The isocentre for the images may be known digitally within the software, following calibration of the on-treatment imaging detector, or it may be inferred by identifying the field edges and (by knowing the field size or position of the MLC leaves) inferring the isocentre position from this. Both methods are performed automatically by the associated software and do not require user intervention regarding the isocentre or field edges. Some centres use a physical reticule placed onto the face of the treatment head that has lead markers accurately aligned with the central axis crosswire of the treatment machine. For this method, the user may need to identify the isocentre position from the opaque markers in each image.

7.5.5.5 2-D Planar Imaging—Bony Anatomy

When bony anatomy is used as the surrogate, the software may automatically identify the bony features on the kV or MV on-treatment images, or it may be done manually by the user. For kV images, the system software is programmed with, at the time of calibration during commissioning and routine QA, the position of the treatment isocentre and can, therefore, compute the geometrical relationship between the identified anatomy and the isocentre. By comparing the anatomy between the planning 2-D DRR and the 2-D on-treatment image, the software can compute the set-up error from the treatment plan (see Figure 7.8). This is the vector between the planned and the on-treatment isocentre positions within the patient. To assess the overall set-up error in three dimensions, a second on-treatment image must be acquired, ideally orthogonally to the first, so the translational discrepancy can be related to longitudinal, lateral, and vertical couch movements for correction.

The bony anatomy in each image is aligned or registered, overlaying one image (e.g. planned) or set of outlined structures over the other (e.g. on-treatment) and the system computes the set-up error from the difference between indicated isocentre positions or field edges. This can be done manually by the user or automatically in some systems. Correction of set-up error could occur immediately, by adjustment of the treatment couch before delivering the full treatment

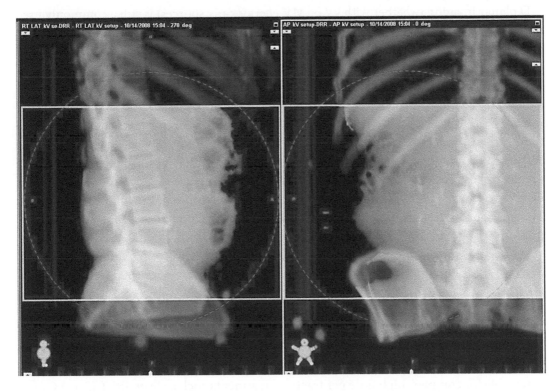

FIGURE 7.8 The computer interface from Varian's on-treatment verification imaging system for 2-D kV-kV orthogonal imaging pairs of the abdomen with colour blend. The left pane shows the right lateral image; the right pane shows the AP view.

Image courtesy of Varian Medical Systems.

fraction (on-line), or computed for an appropriate retrospective correction (off-line) when combined with data for a number of consecutive fractions to calculate any systematic error. Both methods have been discussed in chapter 6. Correcting for systematic and random set-up error components thereby brings the anatomy back into the planned relationship with the treatment isocentre and, therefore, with respect to the high-dose volume of the treatment plan.

7.5.5.6 2-D Planar Imaging—Fiducials

For fiducial markers, the principle is identical. The position and location of the markers are identified on the planning scan and/or the planning DRR. Here, one is relating the position of the markers to the desired treatment isocentre, making the assumption that the relationship of the markers to the target tissues remains constant and that they move as a single, rigid body. On-treatment images are acquired (ideally as an orthogonal pair) if 3-D correction of set-up is required from the 2-D images. These can be either using the MV treatment x-ray beam or a separate kV x-ray source. The position of the fiducial markers are either identified manually on the on-treatment image, or automatically by the software and the two sets of points (planned and on-treatment) aligned.

7.5.5.7 2-D Planar Imaging—Other Technological Points

2-D planar imaging (whether with kV or MV energy x-ray sources) is still a commonly used form of on-treatment imaging for geometrical verification for image guidance in radiotherapy. It is relatively straight forward in its use and, with automatic gantry positioning and couch correction from outside the room, it is very fast and easy for on-line correction of patient set-up. This enables correction of both systematic and random error components on a daily basis with little concomitant dose burden when kV x-rays are used, or when fiducial markers are used with either kV or MV x-rays. There have been generational changes in hardware (Langmack 2001; Herman 2005; Kirby and Glendinning 2006; van Herk 2007) and software that has greatly sped up the on-treatment imaging process. Orthogonal image pairs are possible as either MV-MV, kV-kV or MV-kV. With the use of gold seed fiducials, MV or kV imaging is possible (Mutanga et al. 2012), together with real-time movie loop images available during treatment of static fields for monitoring internal organ or target volume motion (Kirby et al. 1995; Kirby and Glendinning 006; De Los Santos 2013; Zou et al. 2018).

As such, it has been used widely for guiding complex delivery techniques like IMRT and VMAT, for analysing site-specific set-up errors to calculate optimised CTV-PTV margins, and even ensuring their reduction in some clinical sites (Zou et al. 2018), with 5 mm being possible and robust for head and neck patients with 2-D imaging. Certain anatomical changes (in, for example, lung, head and neck, and prostate treatments) can be observed even with only MV 2-D images, which is pertinent to the treatment delivery and dose to the target volume and surrounding OARs (Zou et al. 2018). Intrafractional motion is detectable for complex, high dose per fraction treatments such as liver SBRT (5 to 8 Gy per fraction VMAT treatments), using combinations of kV and MV 2-D imaging during arc treatment delivery. This type of real-time imaging has also highlighted differences between well-established 4-DCT planning methods (using surrogate surface markers) and actual internal target motion during treatment (Zou et al. 2018).

7.5.6 Current Methods With MV and kV X-Rays—3-D Imaging

7.5.6.1 Evolution of 3-D Imaging Dedicated for C-arm Linacs

As we have discussed in the image science sections above, 2-D x-ray imaging is always a projection of the mean attenuation coefficients through the full 3-D anatomy of the patient; it is therefore limited in what anatomical detail is visible in the image. Though highly valuable in what is achievable with 2-D planar imaging, especially in calculating 3-D set-up corrections with some correction possible for rotational errors, it will not reveal further anatomical detail (especially soft tissue) nor be able to analyse out-of-plane rotations robustly. The use of surrogates is also limiting when the soft-tissue target can move independently of the surrogate and one is treating a clinical site where implanting fiducial markers is impractical. There are also schools of thought that value more highly the ability to see in detail the full 3-D anatomy for the position and size of OARs in addition to the target volume—something 2-D imaging with fiducials is unable to achieve.

3-D volumetric on-treatment imaging on the gantry of C-arm linacs was realised with both the advent of better quality imaging (through AMFPI) and the development of filtered back projection algorithms that could be applied to a cone beam (rather than fan beam) of x-rays—cone-beam computed tomography (CBCT). The development of CBCT brought a level of detail into the treatment room that had only been possible in a cursory way with previous EPID technology. Real volumetric data similar to the quality of planning CT was now possible for localisation and on-treatment geometric verification (see Figure 7.9). Although conformal, IMRT and other complex radiotherapy techniques were implemented and guided geometrically without it, 3-D volumetric on-treatment imaging realised IGRT in its modern sense.

Like conventional CT, CBCT is acquired by rotating the x-ray source (either kV or MV) around the patient whilst acquiring image data from a diametrically opposed detector. For conventional CT, this is a row (or rows) of individual detectors and the x-ray beam is collimated into a fan beam; for CBCT, the beam is conical and the detector is an area detector—an AMFPI. Instead of multiple rotations of source and detector around the patient, whilst the couch moves, in CBCT, the couch and patient are stationery and the full 3-D image is acquired with one single rotation of the gantry (either a full 360 degrees, or a partial arc dependent upon clinical site and desired image quality). Attached to the gantry of the linac, the rotation during acquisition is limited to one revolution per minute, by international legislation. It can be achieved using both the MV treatment beam and/or a separate kV x-ray source. Commercially, Elekta and Varian market versions have the latter, attached to the gantry at 90 degrees to the MV treatment beam (see Figure 7.10) and its own EPID. Siemens, before they withdrew from the radiotherapy market, produced a commercial linac that could produce both kV and MV CBCT images, with the kV source and its imaging panel mounted 180 degrees to the MV treatment beam (De Los Santos et al. 2013).

7.5.6.2 MV-Based CBCT

Some of the earliest clinical CBCT images were acquired using MV CBCT (DLS 46–49) and, as such, much initial research was developed examining the possibilities of EPID-based in vivo dosimetry and then, ultimately, dose-guided radiation therapy since both

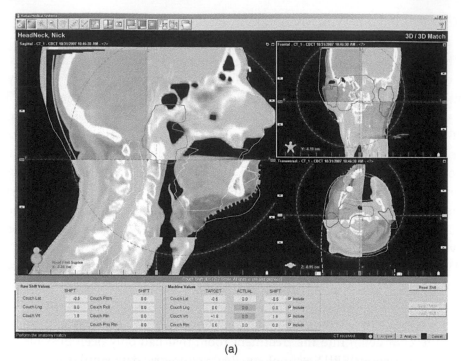

(a)

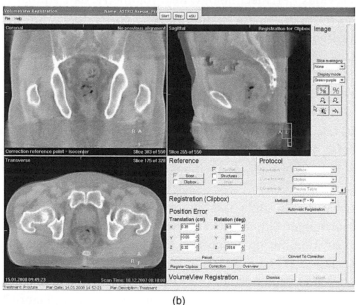

(b)

FIGURE 7.9 The computer interfaces during volumetric cone-beam CT (kV CBCT) showing image matching of the CBCT with planning CT scans for (A) a head and neck patient and (B) a pelvis patient. In (A), the images are split into quadrants showing, alternately, the CT reference images and the on-treatment kV CBCT images (*Image courtesy of Varian Medical Systems*). In (B), a colourwash is used with purple being the reference planning CT dataset and green being the on-treatment kV CBCT dataset.

Image courtesy of Elekta.

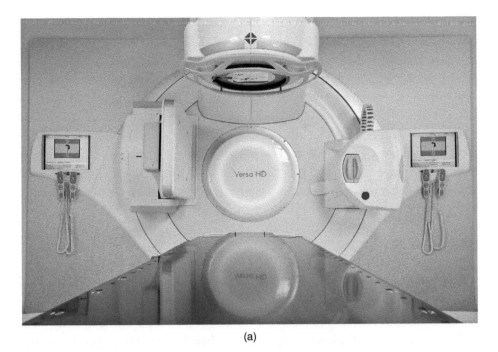

(a)

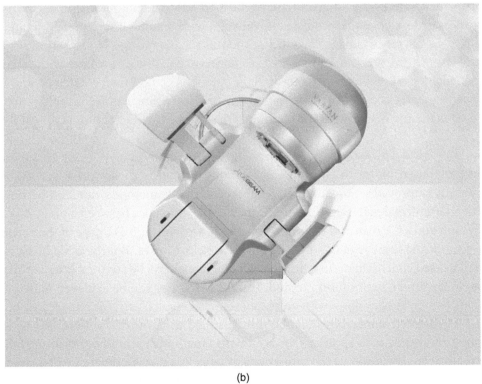

(b)

FIGURE 7.10   Elekta's (A) and Varian's (B) modern C-arm linacs with kV imaging arms and flat-panel detectors clearly visible at 90 degrees to the treatment head.

Image (A) courtesy of Elekta; image (B) courtesy of Varian Medical Systems.

geometric and dosimetric information can be gleaned with the MV treatment beam (Chen et al. 2006). The earliest MV CBCT images showed the value of the improved detail possible with volumetric methods, especially in the head and neck region, and also the value of MV-based imaging with fewer imaging artefacts from implanted markers, dental fillings, and metal prostheses, compared with kV based imaging. However, improved imaging came with the cost of a higher concomitant dose burden (de Los Santos et al. 2013). Some still favour MV-based techniques, since the geometry is identical to that for treatment delivery (treatment and imaging isocentres are the same), with research into beam-line modifications to improve image quality and reduce dose for on-treatment verification. However, the issue of poorer subject contrast for beams not in the kV spectra remains a challenge (Chen et al. 2006; Morrow et al. 2012).

7.5.6.3 kV-Based CBCT

kV-based CBCT is by far the most popular method of volumetric imaging for traditional C-arm linacs, with its image quality and detail that brought a significant shift to on-treatment imaging and set-up correction (Zou et al. 2018). Much research was conducted (and still continues) in reducing the doses involved for kV CBCT to make daily volumetric imaging a possibility with the lowest concomitant dose burden; the key concept being that image quality is necessary for geometric verification only, and not for diagnostic purposes. Software was developed that would perform volume-to-volume image matching and registration; the reference "image" now being the full CT planning 3-D dataset, rather than DRRs. Set-up correction translations are the output of the software together with rotational components for correction, since out-of-plane rotations could now be detected and computed with greater confidence. Localisation accuracy is within 1 mm for registration and 3-D–3-D matching of images to the planning CT scan data (de Los Santos et al. 2013).

7.5.6.4 kV-Based CBCT: Experience, Advantages, and Disadvantages

Volumetric-based on-treatment imaging methods, primarily those which are kV-based, have greatly improved set-up accuracy compared with 2-D methods, not least because the level of anatomical detail is far superior. Consider the difference between a plain 2-D chest radiograph and a CT scan of the thorax; or between an anterior 2-D image of the pelvis and a CT scan of the same region—the advantages are clear to see. But the added cost is that of an increased concomitant dose burden, which must always be factored into justification of the on-treatment verification method. If set-up accuracy can be achieved with 2-D kV methods (which are low dose) with acceptable clinical results, then the added dose burden from 3-D volumetric methods must be fully justified in terms of the potential benefit if the same frequency is to be used. The volumetric methods naturally afford much better soft-tissue detail, even within the limitations of x-ray imaging. Compared to 2-D methods and the use of surrogates, kV CBCT helps to visualise the full 3-D soft-tissue volumes of OARs, as well as the target tissues, making monitoring internal anatomical changes considerably easier and adaptive radiotherapy a distinct possibility (see chapter 10). The visualisation of volumetric soft-tissues using CBCT is unparalleled compared with 2-D planar imaging.

Image quality is naturally inferior to that of conventional fan beam CT used for diagnosis and pretreatment planning because of the additional patient scatter present in the CBCT method (Zou 89). Recent research examines ways of improving upon this through automatic scatter correction algorithms (Wang et al. 2016), which model and predict the scatter within each projection image and use iterative reconstruction to help eliminate some of the noise from the scatter.

Less of an advantage is the increased time required for CBCT acquisition and analysis, for both on-line and off-line correction; 2-D methods are quick and time-efficient in most cases. There is also a lack of real-time 3-D data during treatment delivery to monitor intrafractional motion live, which is possible as we have seen with some 2-D methods. The magnitude of intrafractional motion can be inferred three-dimensionally by acquiring volumetric CBCT data at the start and end of the treatment fraction, but adds to the time and resource burden for each treatment slot.

Daily kV CBCT on-treatment image guidance has been shown to be highly valuable in correcting for set-up errors (systematic and combined systematic/random) and analysing set-up accuracy for clinical sites and techniques, thereby helping to compute and reduce CTV-PTV margins for treatment planning. It has been used for IMRT, VMAT, and SBRT/SRS techniques and in clinical sites of lung/thorax, liver, head and neck, brain, liver, pelvis, and spine (de Los Santos). Since scan acquisition is slow (one to two minutes), the image acquired is an "average" of positional information where internal organ motion (e.g. respiration) is likely (De Los Santos et al. 2013). In terms of margins, for example, prostate margins could be reduced to approximately 2 mm by using kV-based volumetric on-treatment imaging (Morrow et al. 2012, Yartsev and Bauman 2016; McPartlin et al. 2016; Ahunbay et al. 2016). For head and neck patients, margins could be reduced to less than 2.5 mm under volumetric imaging, although weight-loss during the treatment course must be carefully monitored and may necessitate a slightly larger margin (Zou et al. 2018). Being volumetric, tumour response during treatment can also be monitored (De Los Santos et al. 2013).

7.6 ON-TREATMENT IMAGING USING IN-ROOM X-RAY TECHNOLOGY ON C-ARM LINACS

7.6.1 In-room kV Imaging Technologies

Alongside EPID development and on-treatment imaging technology attached to the gantry of the linac, researchers also developed other kV imaging modalities for use within the treatment room. These would bring the advantages of kV imaging, so image quality was essentially the same as 2-D and 3-D imaging used for diagnosis and pretreatment planning, but also afford flexibility in not being attached to the linac gantry, making interfractional imaging a possibility in one particular case. The first development was to introduce a conventional CT scanner into the treatment room, but with an interesting operational development; the second was to bring orthogonal kV x-ray imaging into the room, thus affording stereoscopic 2-D imaging for before, during, and after delivery of a treatment fraction (De Los Santos et al. 2013; Dieterich et al. 2016a). The former is no longer a clinical product, but the latter is still used by Varian/Brainlab in its ExacTrac system and also by Accuray's Cyberknife (Zou et al. 2018; Dieterich et al. 2016a).

7.6.2 CT on Rails

The concept of using the same imaging modality and technology for both pretreatment planning and on-treatment imaging is an elegant one; the best and most robust image matching and registration in the on-treatment setting is likely when the modalities and energies used are identical (i.e. kV to kV, 2-D to 2-D, 3-D to 3-D etc.). This concept was developed by installing a CT scanner within the linac treatment room, with Siemens and Varian being some of the earliest developers of this hybrid, and one fully commercialised by Siemens alone. The CT scanner was identical to that used for pretreatment CT planning, thereby delivering an identical image quality, dose (about 10 to 50 mGy) and localisation accuracy (≤1 mm) (De Los Santos et al. 2013; RCR 2008a). The method of acquisition was identical to pretreatment procedures, except that instead of the couch translating through the bore of the scanner, the scanner was mounted on rails and moved over the patient on the treatment couch of the linac during the scan. The scanner was placed at 90 or 180 degrees to the linac, and the patient on the treatment couch was rotated between a scanning position and a treatment position. In the scanning position, the scanner would move smoothly on rails rigidly mounted into the floor of the room, thus, there was a known and precise geometric relationship between the scanner's isocentre, scanning slice position, and the isocentre of the linac within the room. Imaging quality is therefore identical to the 3-D or 4-D methods used for pretreatment planning scans. Any on-line set-up correction necessary is calculated whilst the couch is rotated from a scanning to a treatment position and then effected on the couch in readiness for treatment.

Initial designs used a standard-bore CT scanner, later developing into using wide-bore, multislice versions (De Los Santos et al. 2013; Zou et al. 2018). With its image quality, it was ideal for monitoring tumour response and plan adaptation during treatment. The configuration has been used for clinical sites such as prostate, lung/thorax, head and neck, spine, and for inter/intrafractional verification, the latter by taking scans at the start and the end of the treatment fraction (De Los Santos et al. 2013). Direct soft-tissue localisation was easily possible within the standard limitations of x-ray imaging, but no worse than the image quality used for target volume/OAR delineation at treatment planning—and without the need for other imaging surrogates. Interobserver variation was observed to be reduced in terms of delineation and identifying differences between target volumes and surrogates. Like kV CBCT, the on-treatment imaging modality and treatment don't share a common isocentre. This required careful calibration and quality assurance (QA) to ensure a robust and accurate relationship between the two, including couch movements and readouts (linear and rotational) (De Los Santos et al. 2013; Dieterich et al. 2016a).

7.6.3 2-D kV Stereoscopic In-room Imaging

The concept of using orthogonal kV imaging pairs was translated to in-room imaging in commercial forms such as ExacTrac (from Varian/Brainlab initially on their Novalis SRS platform and later marketed for other linacs) and Accuray's Cyberknife (Dieterich et al. 2016a; Zou et al. 2018). Its precursor might have been seen as the Real-time Tumour-tracking RT system developed in Japan as a quadruple orthogonal kV on-treatment imaging system for real-time tumour tracking using fiducial markers for gated treatments

(Shirato et al. 2004). Four x-ray tubes and image detectors (image intensifiers) were used so that verification was possible throughout treatment, without being obscured by gantry position. Accuracy was <1 mm, with an associated accuracy for moving targets (up to 40 mm/s) of <1.5 mm whilst monitoring internal anatomy or implanted fiducials under fluoroscopy in 4-D, which naturally could register very high skin doses (>1000 mGy). Clinical sites included lung, liver, prostate, and spinal tumours (De Los Santos et al 2013).

ExacTrac, which is still a commercial product, utilises a pair of 2-D planar kV orthogonal x-ray tube/AMFPI pairs in a cross-fire pattern through the linac isocentre (De Los Santos et al. 2013; Zou et al. 2018). Designed and used mainly on Varian's Novalis system for SRS/SBRT to afford frameless but highly accurate treatments, it can be installed on all linacs. Orthogonal 2-D snapshot images are acquired for initial set-up correction and then intrafractionally at various time intervals to assess and monitor motion. A basic 2-D kV image pair for initial set-up could reflect a concomitant dose of approximately 0.5 mGy, but used intrafractionally, it could produce a much higher concomitant dose for the patient— on the order of tens of mGy (De Los Santos et al. 2013; Ding et al. 2018). Three-dimensional and 6-D set-up correction is possible, depending upon the degrees of freedom available on the couch (De Los Santos et al. 2013). Technically, it can be interfaced to the linac, so that if intrafractional movement is deemed outside predefined tolerances, the treatment beam is automatically interrupted. Combined with nonionising radiation techniques (e.g. IR external markers), it becomes a hybrid on-treatment imaging system for initial set-up and motion monitoring during treatment; the Novalis linac is also equipped with 2-D MV, 2-D kV, and 3-D kV CBCT on-treatment imaging technologies (DLS 26). Calibration and routine QA between the different imaging modalities and the treatment isocentre is vital (De Los Santos et al. 2013; Dieterich et al. 2016a, 2016b).

Clinically, it has been used for intra- and extracranial radiotherapy—sites like the brain, lung, liver, head and neck, and spine—through the on-treatment imaging of bony surrogates or implanted fiducial markers, since soft-tissue is generally difficult to discern from the 2-D kV images (De Los Santos et al. 2013). Initial use was for cranial SRS as a frameless option for treatment delivery, with some studies showing comparable accuracy to frame-based Gamma Knife treatments (Zou et al. 2018). Making use of all its imaging modalities (including those using nonionising radiation), accuracy can be submillimetre. Similarly, Cyberknife technology also uses a stereoscopic kV and nonionising radiation on-treatment imaging system (see chapter 10).

7.7 ON-TREATMENT IMAGING USING NONIONISING RADIATION TECHNOLOGY AROUND TRADITIONAL C ARM LINACS

7.7.1 Rationale

On-treatment imaging technologies that do not use ionising radiation are favored in that there is no contribution to the concomitant dose burden for the patient. This offers the potential for much higher frequency of use, either daily or in real-time, particularly for real-time monitoring and localisation throughout the daily fractional delivery where intrafractional motion is likely to be significant (Bortfeld and Chen 2004, Yan 2010).

There are four broad categories involved:

- Those using MR guidance during treatment delivery (to be discussed in chapter 10)

- Those using ultrasound

- Those using surface-monitoring techniques, either through markers attached to or connected with the patient, by acquiring images directly of the patient's surface

- Those using implanted transponders

Key source references for these technologies include research by Bourland (2012), Meeks et al. (2012), Molloy (2012), de los Santos et al. (2013), D'Ambrosio et al. (2012), and Yartsev and Bauman (2016).

7.7.2 Technologies and Equipment

7.7.2.1 Ultrasound

The use of high-frequency sound for acquiring images (2-D, 3-D and 4-D) of internal structures in or around the target volume has been used for some time, especially for diagnostic purposes. For radiotherapy practice, ultrasound-based reference images can be acquired at pretreatment within the CT simulation suite and/or images registered with structures outlined during virtual simulation and/or computerised treatment planning. On-treatment, the position and orientation of the ultrasound probe/transducer can be linked either optically or mechanically to an in-room detection system (usually using and optical system with infrared cameras) which relates its position to the treatment room isocentre (see Figure 7.11). The probe has infrared markers on it, and, through relevant calibration procedures and software, its position in the room can be detected to at least millimetre accuracy (Mark et al. 2005, Wan et al. 2008). Mechanical methods can also be used through an articulated arm again linking the probe to the treatment isocentre (Mohan et al. 2000, Morr et al. 2002). Daily images can be acquired for each fraction and registered with pretreatment planning target structures immediately prior to treatment when the patient is set-up on the treatment couch. Most often used is an on-line protocol for correcting geometric positioning errors (Lattanzi et al. 1999, Morr et al. 2002, Serago et al. 2002, Langen et al. 2003).

There are a few clear benefits. First, there is no radiation dose associated with this type of on-treatment image guidance. Second, it is noninvasive and provides geometric verification via internal structures around the target volume and with respect to OARs. Also, image acquisition, comparison, analysis, and set-up correction usually takes just a few minutes—accuracy is generally about 3 to 5 mm with reasonably good visualisation of soft tissues and target volumes (Boda-Heggermann et al. 2008, de Los Santos et al. 2013). Plus, all of this is done in a highly cost-effective manner. For patients and treatments where secondary cancer induction is of particular concern, on-treatment imaging in this way with no concomitant dose could be highly desirable (Hricak et al. 2011).

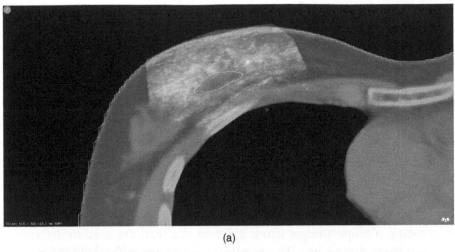

(a)

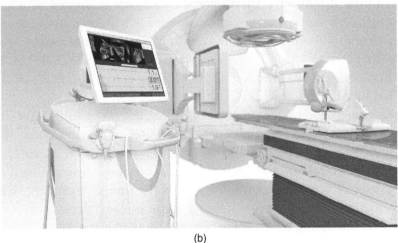

(b)

FIGURE 7.11 (A) An example of an ultrasound on-treatment image matched with the reference-planning CT scans for this right-sided breast patient. A colour wash is used, with the ultrasound image in yellow. (B) shows the Clarity ultrasound system used for trans-perineal on-treatment imaging for geometric verification of the prostate.

Images courtesy of Elekta.

Ultrasound imaging is not without its challenges, though; imaging can be compromised if there isn't an appropriate "acoustic window" for the ultrasonic beam to enter and pass through the body, i.e. minimal bony obstructions are required between the skin's surface and the target volumes (Molloy 2012). Additionally, the pressure of the ultrasound probe can be an issue, in causing deformation of the underlying tissues thereby giving false image information for the clinical set-up during treatment delivery (de Los Santos et al. 2013). Most experience for clinical sites has been obtained with prostate (Boda-Heggemann et al. 2008) and breast (Petersen et al. 2007, Berrang et al. 2009), with potential for future 4-D localisation in real-time for abdomen and prostate sites (Falco et al. 2012, Schlosser et al. 2016, Sihono et al. 2017, Martyn et al. 2017).

7.7.2.2 Surface Methods

Some on-treatment technologies can employ a continuous monitoring of the patient's surface in real time, quickly, and with no concomitant dose burden. This can be either through the use of monitoring external markers or the patient's surface itself (Chen et al. 2005, Bert et al. 2006, Meeks et al. 2012, D'Ambrosio et al. 2012, de Los Santos 2013). Optical tracking can detect marker objects in 3-D space, which can then be related to the patient's external or internal anatomy and to the treatment room isocentre. Optical detection is most common; imaging markers can either emit or reflect infrared light. These reflective objects are then monitored (in real time) by camera systems rigidly mounted in the ceiling, on the walls, or on brackets suspended in the treatment room.

The systems use the principle of photogrammetry, which mathematically calculates the distances between objects in an image by knowing the scale of the image. With stereo-photogrammetry, multiple images (from different cameras) are used to determine object coordinates in three dimensions. These optical systems are calibrated with respect to the room isocentre, have a high spatial resolution, operate in real time, and can resolve objects (through cameras and IR sources mounted rigidly in the treatment rooms) to a fraction of a millimetre (Menke et al. 1994, Bova et al. 1997, Baroni et al. 1998, Kai et al. 1998, Rogus et al. 1999, Kubo et al. 2000, Meeks et al. 2000, Wang et al. 2001, Soete et al. 2002, Yan et al. 2003, Tenn et al. 2005).

Surface tracking uses the same principle of photogrammetry, using light patterns and/or light projected or scanned onto the patient's surface (Bert et al. 2005 and 2006, Schoffel et al. 2007, Brahme et al. 2008, Cervino et al. 2012, D'Ambrosio et al. 2012, Willoughby et al. 2012, Shah et al. 2013a). A single or multiple scanning/camera system can be used to model the surface of the object (the patient) at the time of treatment delivery. Rigid body transformations are used to compute and minimise the differences between planned and what is observed during treatment delivery (Willoughby et al 2012, D'Ambrosio et al. 2012). For accurate, precise, and robust work, the camera and light projection systems must be rigidly mounted to the ceiling or walls of the treatment room.

Both methods make broad assumptions of the relationship between the external markers or the patient's skin surface to the target volume structures identified at pretreatment scanning and planning. Geometric verification is achieved through this assumption or with respect to information taken in the treatment room at the time of first fraction once the patient has been geometrically verified and positioning corrected for set-up error (perhaps using other methods and technologies).

Reported accuracies from phantom and patient studies can be 1 to 2 mm or even sub-millimetre (Meeks et al. 2000, Bert et al. 2005, 2006, Schoffel et al. 2007, D'Ambrosio et al. 2012, de los Santos et al. 2013, Shah et al. 2013a). Comparison with x-ray methods can sometimes be poorer (Zhao et al. 2016). The methods allow for the possibility of thresholding and gating treatments (interrupting the treatment beam if set-up errors are detected and are outside a predefined tolerance) during continuous patient monitoring, and even for verifying the correct patient ID (Wiant et al. 2016). Clinical use has primarily focused on breast and chest-wall treatments (Shah et al. 2013a, Batin et al. 2016, Walston et al. 2017, Zagar et al. 2017), pelvis (Pallotta et al. 2015, Apicella et al. 2016, Zhao et al. 2016,

Krengli et al. 2016), cranial (Manger et al. 2015, Li et al. 2015, Lau et al. 2017), for SBRT pro-cedures (Cervino et al. 2012), and sometimes combined with other nonionising modalities (Krengli et al. 2016, Fattori et al. 2017).

7.7.2.3 Implanted Transponders

These systems allow for real-time, continuous monitoring of target tissues during radio-therapy by means of beacon (wireless) transponders implanted within the target volume. Implantation is similar to that used for fiducial markers for x-ray-based systems. One main system—the Calypso system—uses a mobile emitter/receiver array positioned above the patient during treatment, ceiling mounted IR camera systems (for tracking and monitor-ing the array), and associated computer hardware/software (Balter et al. 2005, D'Ambrosio et al. 2012, de los Santos 2013) (see Figure 7.12).

The position of the detector array is calibrated with respect to the room isocentre through transponders implanted accurately in a phantom, which is set-up in a precise and known location with respect to the treatment room isocentre (Balter et al. 2005, Willoughby et al. 2006, Litzenberg et al. 2007, Santanam et al. 2009, Willoughby et al. 2012).

In clinical use, transponders are implanted pre-CT scanning, and their positions noted with respect to the isocentre of the treatment plan. In the treatment room, patient

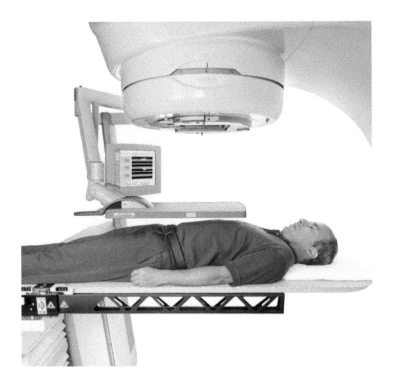

FIGURE 7.12 A photograph of the Varian Calypso on-treatment verification system, which uses inplanted beacon trasnponders for geometric verification on-treatment. The detector array is sus-pended above and close to the patient for detection of the trasnponder's positon.

Image courtesy of Varian Medical Systems.

positioning is adjusted on-line prior to treatment delivery, and then monitored continuously during the fraction with the possibility of interrupting the beam if positional variations are detected outside a predefined tolerance. Clinical sites have mainly involved the prostate with investigational studies on the head and neck, thorax, and GI sites (Parikh et al. 2007, Sandler et al. 2010, Tanyi et al. 2010, D'Ambrosio et al. 2012, Shah et al. 2013b, Lovelock et al. 2015). Localisation accuracy is submillimetre for phantom studies (Balter et al. 2005) and within 1 to 2 mm of methods using ionising radiation in the clinical setting (Willoughby et al. 2006). A novel, single transponder device is also now being investigated clinically (Braide et al. 2018).

CHAPTER 8

Clinical Practice Principles

8.1 INTRODUCTION

Chapter 7 discussed the technological aspects of imaging science and imaging equipment used in the radiotherapy department, both pretreatment and on-treatment. This chapter will begin to discuss clinical applications of imaging equipment with relation to imaging science aspects. Factors influencing choices in equipment will be covered in relation to different treatment techniques and anatomical sites.

8.2 PRETREATMENT IMAGING EQUIPMENT

All patients will have imaging studies completed to aid in diagnosis, following diagnosis to determine stage of the disease, and radiotherapy imaging studies completed to fully determine where exactly the radiotherapy treatment will be delivered. It is of utmost importance during all of these different imaging instances to gain clear and precise information. This may mean that different anatomical areas or different diagnoses of patients will have different image technologies used to give the optimum quality image from which to make informed decisions on best treatment options.

8.2.1 Radioisotope Imaging

Radioisotope imaging gives functional information, but does not have high spatial resolution, therefore, its use in providing accurate structural information is limited. Radioisotope imaging generally provides a 2-D image. As radiotherapy plans generally require volumetric images with a high spatial resolution, radioisotope imaging is more commonly used at diagnosis stage and to determine the staging of the disease, rather than defining an exact target volume for radiotherapy treatment.

8.2.2 PET Imaging and PET/CT

In a similar way to radioisotope imaging, PET images provide functional information. PET alone has a poor spatial resolution and, therefore, cannot give sufficient structural information to guide radiotherapy treatment planning and positioning. PET images cannot be used to define the extent of or to outline target areas or OARs. PET can be readily combined with other imaging modalities that are able to give structural information, such as CT scanners and MR machines. The combination of the two imaging modalities allows the functional information gained from the PET image to be combined with the structural information from the other modality, the most common of which at the moment is CT. PET/CT images are common in diagnosing and planning radiotherapy treatment to areas such as head and neck and thorax. The use of these combined images is rapidly evolving as the benefit of visualising the functionality of a tumour is becoming realised. Radiotherapy has moved away from treating large fields to ensure coverage and toward very conformal treatment fields to reduce side effects associated with irradiating large areas of healthy tissue. The progress now is toward treating only the actively functioning areas within a visible mass. This has implications to many different anatomical areas as well as different diagnoses. Many studies have looked at the delineation of GTV for the radiotherapy planning of lung tumours, especially nonsmall cell lung cancer (NSCLC). These studies have similar findings: The use of PET/CT has improved the GTV definition, showing mainly a reduction in size when PET/CT imaging is used, and a change in the staging of the diagnosis with distant nodal involvement discovered following the PET/CT (De Ruysscher et al 2005, Matsuura et al 2017, Zheng et al 2014, Bradley et al 2012). In all cases, the treatment delivered was more conformal to the actual site of the tumour when PET/CT was used for radiotherapy planning when compared to the use of CT alone. The use of such conformal fields has a great effect on the imaging technique to be used during radiotherapy treatment delivery. If more conformal fields are used for treatment, there will be a smaller margin for error allowed around that field, therefore imaging tolerances and action levels associated with these treatments will be lowered. Consideration must be given to using 3-D volumetric imaging and the possibility of daily image verification of treatment position prior to treatment delivery.

The construction of the PET/CT machine may lead people to believe that it is one machine. However, it is two separate machines within one piece of equipment. Within the housing of the PET/CT scanner is a separate PET scanner and CT machine. Both pieces of equipment work individually, one at a time. For any patient undergoing a PET/CT scan, they will have a CT scan acquired and immediately after, the PET component in the machine will begin and a PET scan will be acquired. There is very little time delay between the two scans. Following acquisition, the two separate scans are fused automatically in the computer software system associated with the PET/CT.

8.2.3 Computed Tomography (CT)—3-D

3-D CT still remains the most common form of pretreatment radiotherapy planning imaging used. The majority of radiotherapy departments will have one or a number of dedicated radiotherapy planning CT scanners within the department used solely for the planning of radiotherapy treatments. Regardless of the image quality produced, CT images are required for dosimetric reasons, which have been discussed in the previous chapter. For this chapter, we will focus on the imaging aspect of CT machines. CT images have a high spatial resolution, good subject contrast resolution and a high signal-to-noise ratio. Although some soft-tissue structures can be difficult to distinguish from one another due to similar attenuation, these factors allow the radiotherapy planner to be able to distinguish soft-tissue masses from bony tissue and surrounding air in order to delineate target volumes. As discussed previously, CT scans can be combined with PET data to produce an image including structure and function of an organ or tissue. PET data would not be usable for radiotherapy planning without the structural and more precise positional information gained from the CT scan.

A CT scan can be acquired in a very short length of time and the use of wide-bore scanners has allowed patients to be scanned in their treatment position. Both of these factors make CT scanners ideal equipment for radiotherapy planning. The speed of image acquisition reduces the time during which the patient may move, either consciously or subconsciously. A lengthy scan, using equipment that is not specifically designed for comfort, may encourage the patient to move, and any small movement can cause distortion and artefacts on an image. Reducing the movements made by a patient can reduce the number of these artefacts. Since we cannot, using standard CT methods, control patients' breathing movements, a faster acquisition of the image limits the impact of this motion. It is also important for radiotherapy planning that the patient be in the exact position used at treatment delivery. This allows tattoo or other permanent reference marks to be placed on to the patient's skin at the time of the scan. The position of this mark will be highlighted on the relevant scan slice, and all treatment plan parameters, moves to isocentre etc. will be planned from these marks.

The data acquired from the CT scan can be sent, directly to the treatment planning system for the plan preparation process to begin. Having dedicated CT scanners within the radiotherapy department aids this transfer as the data does not have to be transferred between departments or even between hospitals. CT data has enough information attached to create the treatment plan and the necessary reference images. The imaging data can be used to produce DRRs, DCRs, or exported directly to the treatment machine as a full CT dataset, this full dataset is required as a reference image for cone-beam CT scans.

8.2.4 Computed Tomography (CT)—4-D

A limitation of CT imaging can be artefacts produced due to a patient's normal breathing motion. This problem becomes more of an issue when aiming to visualise a tumour affected by respiratory motion. As well as problems arising with lung tumours, breathing movements affect other areas such as pancreas and liver. Whilst taking a fast CT scan, the tumour is likely to be imaged in a variety of positions. The CT scan image is taken using image slices, one radiation source, and a detector array moved quickly around the patient. During the scan, due to breathing motion, the tumour will move slightly over time depending on the patient's position in the breathing cycle, i.e. at maximum inhalation, the tumour may be pushed to an inferior position and during maximum exhalation, the tumour may move back to a more superior position. See Figure 8.1.

As the radiation source passes around the patient, it is easy for the movement of the tumour at the point of the image being taken to cause a distortion in the shape of the tumour on the image. Instead of the tumour being visualised at its correct dimensions, it may appear to be either elongated or compressed, depending on the position

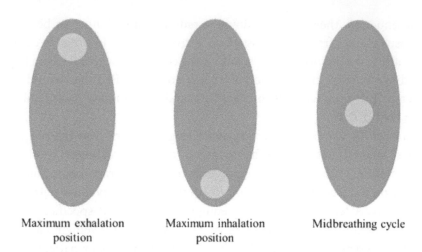

Maximum exhalation
position

Maximum inhalation
position

Midbreathing cycle

FIGURE 8.1 Possible tumour position throughout a normal free-breathing cycle.

of the tumour and the direction of its movement when the radiation passed through it. Images like those represented in Figure 8.2 will be seen instead of a tumour's correct shape and size.

This is not ideal for radiotherapy planning. Instead of being able to outline the exact GTV, an extended volume must be used. This volume will cover the tumour position from maximum inhalation to maximum exhalation. This is required to ensure that, wherever the tumour is during the breathing cycle, the full treatment dose is delivered to the target volume. This volume is known as the internal target volume (ITV), which is much larger than a standard PTV.

ITV can be defined as the CTV with a surrounding 3-D margin to account for all internal uncertainties, including that of organ motion. The ITV does not account for set-up errors, only internal changes of the target volume (Stroom and Heijmen 2002, Aaltonen et al 1997).

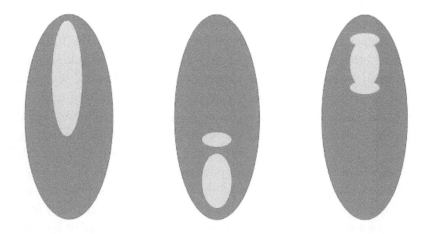

FIGURE 8.2 Possible appearance of tumour during free-breathing 3-D scan.

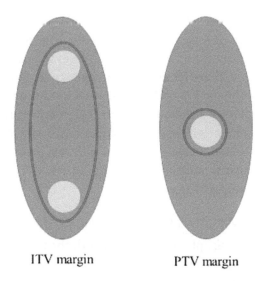

ITV margin   PTV margin

FIGURE 8.3   Examples of possible ITV and PTV margins.

Figure 8.3 shows the difference between the margins required for ITV treatment, if the patient breathes normally, compared to the PTV required if we could image and treat the tumour without breathing affecting its motion. For the treatment margin to be accurately positioned, scans can be acquired using a 4-D technique as described in chapter 7. This allows the tumour to be fully visualised at different points in the breathing cycle without any distortion of its size. It will then be the decision of the treatment planners, based on available equipment and patient need, whether to treat the patient using an ITV or to use a treatment technique, such as gating, to only treat the tumour when it is in a known position. For on-treatment imaging purposes, is it important to know how the tumour will be treated and the margin to be used. If an ITV is used, this is the comparison for the on-treatment images, the patient will be imaged using 3-D in free breathing because it is important to visualise the tumour extent. The surrounding bony structures, visible on 2-D imaging, will not be representative of tumour position. If the patient is to be treated with a smaller target volume using gating equipment, any on-treatment images must be taken in the same position as treatment, i.e. at maximum inhalation. Again it will be important to use 3-D imaging to visualise the tumour itself.

8.2.5 MR Imaging

The use of MR is increasing in radiotherapy. The method of image production is very different from other equipment discussed so far, as it does not involve ionising radiation and so can be used more readily. As there are no dose concerns associated with MR, a patient is more likely to be scheduled for MR imaging, and these can be used in diagnosis, throughout treatment to monitor treatment effects, and after treatment to establish response and monitor late effects. As MR images are produced depending on the response to a magnetic field by individual tissue types (discussed in chapter 7), boundaries between different tissues are much easier to visualise, which leads to a more precise delineation of targets and surrounding organs when compared to CT images.

More limitations of MR include the associated distortion of the image, again discussed in chapter 7, and the length of time the image takes to acquire. Acquiring an MR image takes considerably more time than a CT image. This gives the patient a much longer time-frame in which to move. Any movements during image acquisition will cause distortion or artefacts, and will reduce the image's ability to be used in radiotherapy planning. Suitable immobilisation equipment must also be used for the MR scanner, and it cannot be assumed that all radiotherapy equipment will be either compatible or safe to be used within the MR room. Due to the powerful static magnet involved with this machine, any equipment containing ferromagnetic elements must not be taken in to the MR room. All immobilisation equipment must be MR-compatible for safe use.

8.3 ON-TREATMENT IMAGING ON TRADITIONAL C-ARM LINACS

There are a number of imaging options associated with the modern Linac, as discussed in chapters 2 and 7. Modern linacs have two imaging panels attached, one opposite the MV treatment head to allow MV images of the delivered treatment field to be acquired and a second mounted at 90 degrees to the treatment head opposite a kV radiation source. Both the MV panel and the kV system are capable of acquiring both planar 2-D images and volumetric 3-D scan images; the availability of each image type will depend on the linac system installed. Some systems use MV methods to acquire CBCT, whilst others use only the kV system.

When considering the imaging technique to be used, it is important to consider how the patient is placed in the treatment position and what is included in the target volume. If the patient is positioned using only permanent skin marks, on-treatment imaging is vital as it cannot be assumed that the skin marks will always be in the same position relative to internal anatomy, this has been recognised for many years (Armstrong 1998, Van Herk 2004, Litzenberg et al 2005).

For many patients, the permanent skin marks serve as a basis to begin the correct positioning for treatment, for example, patients with fiducial marker implantation. The final decision on correct treatment placement will be based on the position of the fiducial markers and not the external skin marks. For these patients, on-treatment imaging will be used daily, prior to treatment delivery. It is important to be aware of the fiducial marker used and which energy of radiation may be required to visualise it. Another example of skin marks being used as only a guide for treatment position would be treatments involving margins tightly conformed to soft tissue. For this patient group, volumetric imaging will be required as it will be essential to verify treatment position based on soft-tissue anatomy.

8.3.1 2-D MV Imaging

Acquiring images using the MV image panel and the treatment delivery head can vary in use. It is the only image acquisition method that allows the reviewer to visualise the exact treatment field, as delivered. This is due to the image using radiation dose from the actual treatment field to acquire the image. As there is no MLC capability fitted to the kV source, no kV on-treatment imager can replicate the treatment field parameters. Visualising the actual treatment field can be beneficial when delivering treatment fields close to an organ at risk, the MLC shielding for that organ may need to be reviewed directly to ensure the organ is fully protected. MV on-treatment imaging can be very useful for treatment sites such as bone metastases or breast treatments; the lower subject contrast experienced when using MV energies does not pose a big limitation for these areas as the materials that make up the image are bone, air, and soft tissue. There is a large difference in attenuation, as discussed in chapter 7, between bone and air (found in the lungs and outside of the patient), and between bone and soft tissue. The field arrangement for a whole breast treatment also allows the use of MV treatment images, as the arrangement commonly used is a tangential air with angled fields across the chest wall, the radiation beam does not pass through a number of different anatomical structures before forming a 2-D image. This means that the attenuation of the beam is affected by limited anatomy, resulting in an image that is clearer than if the radiation had passed through the full thickness of a patient. This gives a clear image of the position of the bony structure, the ribs, the lung position, and the outer edge of the breast tissue. Using MV imaging also allows the image reviewer to visualise any MLC configuration used to minimise treatment delivery to the lung. This method of imaging delivers no extra radiation dose to the patient as the treatment field is used, so no justification of images is required; this imaging technique can be repeated as often as necessary without the need for explanation of the dose delivered through imaging.

Radiation fields used to treat whole breast or bone metastases target volumes are often quite large in modern radiotherapy terms. The use of smaller radiation fields conformed to small treatment volumes, such as lung tumours or prostate treatments, do not allow visualisation of enough anatomy to make an informed decision on field placement with MV treatment images. If MV imaging is necessary, possibly due to equipment availability, it is necessary to increase these conformal treatment field sizes to produce an imaging field large enough to evaluate treatment position relative to anatomy. These fields can be referred to as imaging fields, isocentre fields, isocheck fields, or other similar descriptions. As the parameter of these fields no longer matches the treatment field, it is not possible to use a portion of the treatment dose to form the image, therefore, the patient will receive extra radiation dose from imaging. As MV energies are being used, the acquiring of MV imaging fields is the largest dose associated with on-treatment imaging. This dose must be considered; if possible, kV energies should be used to reduce this dose, the size of the MV imaging field should be considered, and the field must be large enough to allow accurate analysis of the treatment position relative to anatomy (but not too large that healthy tissues are receiving radiation unnecessarily).

Images formed using MV energies will only allow treatment position to be evaluated against bony structures. Soft tissue is not visible at this energy. MV imaging, regardless of the imaging field size, can only be used if the bony anatomy is a suitable surrogate for the target volume or if it is the target anatomy, as is the case for bony metastases. When treating a target volume in a soft-tissue organ, as most are, the use of bony anatomy as a surrogate for position of the target is not generally accurate. The position of radiotherapy treatment fields to the prostate, evaluated using bony match, is no longer the method of choice. Research has proved that the prostate can move independently of the pelvis bones that were previously used as a surrogate to evaluate treatment placement (Crook et al 1995,

Dufton et al 2012, Moreau et al 2017). These studies involved the use of fiducial markers implanted in to the prostate and visualised using MV or kV on-treatment imaging to determine the position of the prostate itself. All studies showed a significant movement of the prostate and, therefore, the treatment target that would not have been identified using MV imaging and bony anatomy as a surrogate.

8.3.2 2-D kV Images

The largest benefit of using kV imaging to verify the position of radiation fields is the lower dose delivered to the patient when acquiring these images. As a patient treated with a radical intent is likely to be scheduled for many weeks of radiation treatment, this will involve many on-treatment images to be acquired. The dose received from these images can begin to be significant, especially if tissues outside of the treatment target are included. For this reason it is important to investigate the use of kV imaging as standard wherever possible. Due to improved spatial resolution and subject contrast seen at the lower energies, a clearer image is produced when using kV energies compared to MV. For these reasons, kV imaging equipment is a standard feature on modern C-arm linacs. As discussed earlier, kV equipment does not include MLC equipment and capability, therefore the exact treatment parameters cannot be replicated. If an enlarged imaging field is to be used rather than the planned treatment filed, the use of kV energy will reduce the dose delivered to surrounding healthy tissue and minimise tissue damage caused by radiation from imaging.

As the kV imaging equipment is mounted to the linac gantry separately from the treatment head, it is possible to take a combination of MV and kV images. With the linac gantry remaining at 0 degrees, an MV anterior image can be acquired, directly followed by a kV lateral image. Both of the orthogonal images are acquired in a very short time without moving any parts of the linac gantry. This is the quickest acquisition of on-treatment images, but will give a larger radiation dose to the patient than the use of a kV orthogonal pair, which would involve gantry rotation through 90 degrees.

The use of 2-D kV images for analysis is similar to that for MV images. Regardless of the energy of the radiation used to acquire an image, if that image is 2-D, it will have similar limitations as both energies will produce a 2-D (flat) image of a 3-D structure. Both images MV and kV will allow the reviewer to use bony anatomy in the image analysis. A clearer images comes from kV energies, allowing greater distinction of soft tissue, i.e. if reviewing an on-treatment image of a chest treatment, a large lung collapse will be identifiable on kV images, but volumetric analysis is not possible until a 3-D image is acquired.

MV and kV planar imaging both use similar reference images, either a DRR or DCR produced during the pretreatment plan production stage.

8.3.3 Cone-Beam CT (CBCT)

The term CBCT was introduced in chapter 2. The title "cone beam" is due to the shape of the radiation beam used to acquire the image. The main difference between a CBCT and images produced using a specific CT machine is the patient does not move through the radiation field during a CBCT. The patient is static and the single radiation source and corresponding detector rotate around the patient with a maximum rotation of approximately

365 degrees, just beyond one full rotation of the gantry. Due to this, a wide radiation field is required to produce the 3-D image. The static position of the patient limits the scan length achievable with CBCT. During a CT scan acquired using CT equipment, there is no limit on scan length; in order to image a large volume of the patient, the patient will be moved through the bore of the CT scanner until that scan length has been reached. With CBCT, the scan length is set by the maximum field size achievable on that equipment, the patient cannot be moved through the machine to acquire a longer scan. The centre of the CBCT scan will generally coincide with the treatment isocentre. Due to the limitations of scan length, this may need to be altered. If in the treatment field positioning the isocentre of the machine does not correspond to the centre of the treatment volume, the CBCT scan centre will need to be off-set to visualise the target volume and anatomy to be treated.

CBCT is the only, C-arm linac-based, method of 3-D volumetric imaging. It enables radiotherapy staff to visualise soft tissue and the whole volume of the patient receiving radiation, rather than viewing 2-D information in one plane as with planar imaging. CBCTs can be acquired using MV or kV energies, this decision lies with the choice of manufacturer and equipment purchased, it cannot be selected at the point of imaging. The benefits and limitations of CBCT at the different energies have been discussed in chapter 7. The most popular choice of energy for CBCT acquisition is kV, due to both the increased image quality and lower radiation dose received by the patient than MV CBCT. Although kV energy CBCTs are preferred, they do still deliver a significant radiation dose to the patient, this dose must be justified and consideration must be given to lower dose imaging methods before deciding on the use of CBCT. Despite the associated dose, CBCT has increased in its use as a main method to verify treatment placement. The dose burden from imaging is justified as the ability to visualise target and surrounding structures has allowed for a decrease in CTV-PTV margins, thereby reducing the dose delivered to surrounding healthy tissue during treatment delivery (Wu 2011, Martins 2016, van Nunen 2017).

The visualisation of the full volume of anatomy within the treatment area has led to treating staff now being able to correct previously unnoticed treatment placement discrepancies; examples of these include, rotation for head and neck treatments (Wu 2011 and Li 2008) and alteration of treatment based on discovered lung inhomogeneities (Moller 2014, Kwint 2014). Rotations in anatomy, such as head and neck regions, is not always visible using 2-D imaging techniques, if the rotation is visible, it is often difficult to quantify without the use of volumetric imaging techniques. Similarly, alterations or deteriorations in lung health are not always visible when using planar images. Both of these issues can be common in patients, and CBCT has allowed the treatment position for these patients to be more fully evaluated and decisions to alter treatment to be made based on full volumetric analysis.

CBCT can take up to two minutes to acquire. Due to the complexity and amount of information gained, it can take quite a time for treating staff to fully evaluate the image and make an informed decision on treatment placement before radiation delivery. These factors together can mean an extended time for the patient on the treatment couch and in the treatment position. This can be a limitation of this method of imaging; if the patient is unable to maintain the treatment position, the benefit of CBCT is greatly reduced.

The image analysis may be able to give precise information as to corrections needed in treatment field placement, but if the patient has moved and changed position whilst the treating staff were analysing the image, the process is void. Should the patient move, causing the CBCT acquired to be of no use, the imaging procedure and therefore the dose delivered to the patient, will need to be repeated.

As discussed in chapter 4, training of staff in order for them to gain competence in CBCT analysis must be considered prior to implementing a CBCT protocol for treatment verification.

8.4 CASE EXAMPLES

Here is a treatment case example following the discussion regarding MV and kV planar imaging compared with volumetric imaging is prostate radiotherapy.

The treatment of prostate tumours has become much more conformal in recent years, with the issue of treatment side effects to the rectum and bladder increasing in importance. As the PTV has become smaller and all treatment margins around the prostate have reduced, there has been an increased need for treatment placement verification. As previously discussed, external skin marks and bony anatomy are not suitable surrogates for prostate position as it readily moves during radiotherapy treatment independently of the external skin and bony anatomy (Moreau et al 2017). It is important to visualise the prostate's exact position to reduce the chance of a geographic miss. Two of the main imaging techniques used for prostate treatment verification are CBCT and the use of implanted fiducial markers.

CBCT for verification of prostate position: CBCT has the benefit of full volumetric analysis. Using this method of imaging, it is possible to review the size and state of both the bowel and bladder, both of which are organs at risk in close proximity to the prostate. A portion of both of these organs are very often included in the target volume. CBCT gives the image reviewer much more information for anatomy in the whole treatment volume than 2-D imaging with fiducials, (Moseley 2007).

2-D planar imaging with fiducial marker match: The use of fiducial markers for prostate position verification is a quicker and simpler method of imaging than CBCT. The acquisition time is less and the more simplified method of image match can make analysis quicker, resulting in a much quicker overall treatment time for the patient (Chung 2004).

There have been many studies investigating the use of volumetric imaging and planar imaging using fiducial markers for prostate verification, most of which agree that both methods produce similar results in terms of precision of treatment placement (Barney 2011, Deegan 2015, Das 2013). The difference in opinion is not regarding accuracy of verification, but issues such as interoperator variability and the ability to view surrounding tissues. Planar imaging using fiducials is reported to give less variation between operators and greater intraoperator reproducibility (Goff 2017).

One important consideration must be remembered when deciding a departmental imaging protocol for prostate patients: The dose is larger for CBCT acquisition compared to kV 2-D imaging.

8.5 ON-TREATMENT IMAGING USING IN-ROOM X-RAY TECHNOLOGY AROUND TRADITIONAL C-ARM LINACS

This section will discuss the use of x-ray imaging equipment within the treatment room that is not attached to the treatment gantry as previous equipment discussed has been.

8.5.1 "CT-On-rails"

We have discussed CBCT, its use and its limitations. The benefit of CBCT is the ability to visualise volumetric data for the entire treatment region; the limitations of CBCT lie in its technology and equipment design. The image quality is superior to planar imaging, but lacking when comparing diagnostic equipment. The limitations of scan length can cause an issue if the treatment volume is greater or if the treatment centre is off-set from the treatment isocentre used. These limitations can be overcome with the use of a full working, diagnostic-quality CT machine within the radiotherapy treatment room. This arrangement is commonly described as "CT-on-rails" the technology of which is described in chapter 7. The use of the CT scanner in the treatment room, removes the limitations associated with CBCT whilst bringing the benefits of full volumetric imaging to the radiotherapy treatment. With the CT equipment placed in the treatment room, 180 degrees to the linac head, i.e. opposite the treatment linac. The patient is scanned in treatment position, including all immobilisation equipment necessary, which requires use of a wide-bore CT scanner. As movement of the patient may jeopardise the treatment position, it is the CT scanner that moves around the patient rather than the patient who passes through the bore of the scanner, hence the term "CT on rails." Following image acquisition and analysis, the patient is then treated without need to transfer to a different couch. The process is relatively simple and quick to complete. There are, however, limitations to this system, one of which is the cost. The CT and the treatment linac are two separate pieces of equipment, each must be purchased, installed, and quality-assured. Even though the imaging and then treatment of the patient are completed whilst the patient remains on the same couch, it is necessary to move the patient toward the treatment machine following the CT scan. Also, there is still chance and opportunity for the patient's position to change between acquiring the scan and beginning treatment delivery. If the CT scanner is positioned at 90 degrees to the treatment machine, moving the patient in this direction will give more chance for the patient to move out of treatment position; the 90-degree rotation could cause the patient's position to become unstable. The other concern is equipment stability, i.e. stability of the treatment couch during these movements; this will need to form part of the regular quality control checks for this system.

8.5.2 Stereoscopic In-room Imaging

The benefits of 2-D kV imaging has been discussed, however, one aspect to all of the imaging described so far has yet to be mentioned. Due to the positioning of the equipment used for on-treatment imaging, it has so far only been possible to acquire images of the treatment set-up and position of treatment fields if the treatment couch rotation is kept at 0 degrees. Also, the angle at which the image is acquired will be determined by the angle of the treatment gantry. These issues are addressed by the in-room stereoscopic imaging systems, such as ExacTrac (Varian/Brainlab). For this system, the kV tubes are in the floor and not attached to the treatment gantry, the corresponding kV detectors are ceiling mounted opposite the floor kV tubes, and the treatment gantry is likely to have imaging equipment fitted also, both kV and MV. This allows images to be taken at oblique angles in a cross pattern through the treatment isocentre. The image produced by the kV tube to the right-hand side floor of the treatment gantry is acquired by the detector mounted to the ceiling on the left-hand side of the gantry and vice versa. The floor mounted kV tubes are able to acquire an image with the treatment couch and gantry in the exact treating position. This will allow an image to be evaluated and treatment to begin immediately following that registration, there is no requirement to move the treatment machine or the couch rotation to the correct treating position. This is valuable as not all treatment fields are delivered with the couch rotation at 0 degrees. CBCT has the advantage over kV planar imaging of allowing visualisation of soft tissue, this remains true when using stereoscopic imaging as with ExacTrac (Clemente 2013) and is still the main limitation of the system—only planar imaging is possible. Studies looking at the accuracy of the ExacTrac system, have reported similar results in terms of accuracy when compared to CBCT, mainly during its use for bone matching cranial treatment fields (Jin et al 2008, Ma et al 2009, Ackerly et al 2011).

Following evaluation of CBCT, it may be necessary, depending on treatment angle, to move the treatment couch through a large degree of rotation. It is now not possible to evaluate the treatment field placement prior to delivery as the couch is no longer at 0 degrees— only a stereoscopic system will allow this. When using CBCT to verify treatment position, we have to assume there is no movement to the isocentre, either by treatment couch uncertainty or patient movement, during any couch rotation made. Effects and degree of treatment couch movement when rotating the isocentre will be evaluated during regular equipment quality assurance, patient movement during couch rotation is addressed by the use of immobilisation equipment. Although the issue of patient movement when rotating the couch is controlled as much as possible, it is not possible to evaluate the actual movement experienced when rotating the couch unless stereoscopic imaging is used. The positioning of the kV tube and detector also allows images to be acquired during treatment beam delivery, this is not possible when using linac-based imaging equipment.

The use of a stereoscopic imaging system, if combined with a six-degrees of freedom couch, allows correction of the treatment position to include the normal 3 planes (superior/inferior, right/left, anterior/posterior) and also in three rotational planes (roll, pitch, yaw). Together, correcting a treatment position in each of these six planes, gives an increased level of accuracy as the degree of rotation in set-up can increase the error when using 3-D image comparison (Jin et al 2008).

A limitation of using this stereoscopic imaging equipment is the introduction of another isocentre into the treatment. Regular checks are made to monitor the exact correspondence between all imaging isocentres and the treatment isocentre—the stereoscopic imager isocentre must be included in these checks and it must be featured in the regular quality assurance programme.

8.6 ON-TREATMENT IMAGING USING NONIONISING RADIATION TECHNOLOGY AROUND TRADITIONAL C-ARM LINACS

The use of nonionising radiation imaging methods has one main advantage over other methods in that the patient does not receive any increased radiation dose through imaging. These methods of imaging can be used frequently with no dose justification required.

The main nonionising radiation imaging methods, currently and widely available are:

- Ultrasound

- Surface tracking methods

- Internal transponders

- MR guided radiotherapy (will be discussed in later chapters)

8.6.1 Ultrasound

The use of ultrasound for verification of radiotherapy treatment field position is especially beneficial when used to visualise soft-tissue anatomy. It is a unique way of imaging patients and has very different physics principles involved in the production of a usable image. Ultrasound has been used specifically in verification of prostate position and localisation of breast treatment fields.

8.6.2 Case Examples

8.6.2.1 Prostate Verification

When treating the prostate gland with radiotherapy, it is important, as it is with other organs, to minimise dose to surrounding tissues. The organs at risk around a prostate are the bladder, superiorly to the prostate and the rectum, posteriorly to the prostate. When full, the bladder creates a more favourable sound path than that of the prostate; the rectum also gives a very different sound path and due to this, the boundaries between the prostate and bladder and prostate and rectum are easily identifiable. For this reason, ultrasound, is viewed as a practical alternative to the use of fiducial markers. The main benefits of using ultrasound rather than fiducial markers are: No radiation dose to the patient, the fiducial markers require either kV or MV on-treatment images to be acquired, and with the use of ultrasound, the patient does not require an invasive procedure in order to place the markers within the prostate. Modern ultrasound equipment can be used during radiotherapy treatment to monitor and track any internal changes in prostate position. The ultrasound system can work together with the treatment delivery system to enable automatic shut-off of the treatment beam if the target area, i.e. the prostate, moves out of a predefined tolerance position (Clarity Elekta 2011). Without this level of imaging, it is possible that, during treatment, the prostate position will alter and possibly compromise the dose received by the target. Other imaging equipment capable of this motion monitoring will be discussed later in this chapter.

8.6.2.2 Breast Localisation

For similar reasons that ultrasound is beneficial to prostate position verification, it can be used to localise and verify the position of cavities produced in breast tissue following lumpectomy. These cavities often form the target volume for radiotherapy treatment if partial breast irradiation fields are being used, or for treatment fields planned, to give an extra dose to the tumour bed (boost fields) are used. It is essential for these breast treatments that the cavity to be treated is visible and can be used to correctly place the treatment fields. As the breast cavity becomes filled with fluid, it creates a different sound path than the surrounding breast tissue, meaning ultrasound is an effective way to visualise these cavities. The size and shape of these breast cavities can vary greatly as treatment proceeds. These changes can be due to a number of reasons: Post-operative changes can still occur after the patient has begun radiotherapy treatment, oedema can occur, and the breast area is affected by respiratory motion that will require the cavity to be localised prior to any treatment delivery. The use of ultrasound for this localisation can reduce the radiation dose to the surrounding tissue, minimising risk of secondary tumours and reducing cosmetic affect for the patient (Fontanarosa et al 2015). CT scans can be used for the planning of these breast fields, however, due to possible tissue oedema around any breast cavity, the CT image can give results that are less clear as to which area is breast tissue and which is the cavity. Ultrasound images give a much clearer representation of each area in the breast (Berrang et al 2009). The use of CBCT for on-treatment imaging can be a difficult task to perform for breast patients, there is the consideration of dose to the area as well as the logistical consideration of if the movement of the machine around the patient is possible. Another common method used to visualise the breast cavity involves the use of surgical clips implanted during the surgical lumpectomy. These clips, once inserted, will give a representation of the breast cavity but are viewed via the use of kV imaging energies. In terms of dose received by the patient, this would leave ultrasound to be favourable as no ionising radiation is used.

8.6.2.3 Limitations of Ultrasound in Radiotherapy

Interoperator variability is the variation in results obtained between two operators performing the same task; intraoperator variability is the difference in results obtained by one operator performing the same task a number of times. Ultrasound has been shown to have both a high inter- and intraoperator variability (Van de Meer et al 2013, Fontanarosa et al 2015, Johnston et al 2008). The reduction of inter- and intraoperator variability in ultrasound use may be possible with increased training and experience of the system (Johnston et al 2008, Robinson et al 2012). Training on use of ultrasound in radiotherapy may prove to be a limiting factor for many departments as the cost in both monetary terms and in terms of staff time may well prove too high.

When using ultrasound imaging, the body contour of the treatment area is not visible—this could prove to be a limitation, as identification of internal structures may become confused (Fontanarosa et al 2015).

The method of acquiring ultrasound images is via a probe placed externally on the patient's skin surface. The pressure exerted on to the probe alters the image quality since sound waves travel differently through air than soft tissue. The presence of air in the image can cause the sound waves to be reflected, making it difficult to visualise structures behind the air. Fluid, however, is a very good transmitter of sound waves, resulting in very little reflection of the sound back toward the probe. Increasing the pressure on the probe will minimise any air trapped between the probe and the patient's skin surface, however, increased pressure exerted on the probe can lead to displacement of underlying anatomy (Dobler et al 2006, Van de Meer et al 2013, Scarbrough et al 2006). The use of perineal ultrasound for prostate localisation can remove some issues associated with anatomy displacement. Using abdominal ultrasound requires the ultra sound probe to be removed during CT acquisition used for treatment planning and during treatment delivery. Acquiring the planning scan without the probe in place will mean the anatomy is not affected by the probe, but using the probe to localise the anatomy before treatment delivery will alter the position of the prostate. This will make the CT planning scan and the on-treatment image not effectively comparable to each other. The perineal ultrasound probe can be placed prior to CT acquisition and treatment delivery and remain in place throughout each procedure. This will limit the differences in anatomy displacement between CT reference scan and on-treatment imaging. The perineal probe can remain in place throughout treatment delivery, ensuring organ motion is monitored throughout treatment (Li et al 2017).

Care should be taken with the use of ultrasound for treatment verification, adequate training for staff must be in place.

8.6.3 Surface Tracking

Surface-guided radiotherapy (SGRT) is the use of camera systems to monitor external movements of the patient's surface. This can be achieved either by imaging the patient's actual skin surface or with the use of externally placed markers that are monitored for movement. Surface tracking is noninvasive, does not use ionising radiation, and does not displace anatomy as there is no patient contact. There are a number of commercially available systems, including; AlignRT (Vision RT, London, UK) and Catalyst (C-Rad, Upsalla, Sweden). Systems use two or three ceiling- or wall-mounted light projectors and cameras to detect light reflected from the patient's surface. A reference image of the patient's surface is taken at the time of the planning scan. These images are then exported to the surface-tracking software. An image of the patient's surface is taken daily and compared to the reference image. The light reflected by the patient's surface and detected in the surface-tracking software is continual—any motion during treatment will be detected and the treatment delivery halted should that motion take the patient's position out of agreed and set tolerances.

This surface tracking system can be used for any anatomical site, but most work within radiotherapy has concentrated on breast irradiation, either for partial breast treatment or treatment whilst in deep inspiration breath hold (DIBH), or the use of surface tracking to improve accuracy and patient comfort during cranial stereotactic radiosurgery (SRS). It is hoped that by using a surface-tracking system there can be a reduction in on-treatment imaging required and, therefore, a reduction in radiation dose delivered to patients during treatment (Hoisak and Pawlicki 2018). The accuracy of the surface tracking systems have been discussed in chapter 7 and hold some answers to the value of this system clinically.

It may seem an unusual concept to radiotherapy treating staff that they require computer software to monitor if the patient moves as it is an inherent part of treating patients that they are monitored throughout. With accuracies of as small as 1 to 2 mm or less being reported (Meeks et al 2000, D'Ambrosio et al 2012, Shah et al 2013), it becomes apparent that treating staff could not detect movement in the patient's position to that degree of accuracy. Using on-treatment imaging prior to treatment delivery will ensure that the patient's position and, therefore, the treatment field position are within acceptable tolerances; if the patient's position is not accurately monitored throughout treatment, there is no guarantee that all of the treatment will be delivered to the target volume with the same level of accuracy.

8.6.4 Case Examples

8.6.4.1 SRS

Stereotactic radiosurgery is the treatment of small intracranial tumours with a high dose and steep dose gradient to minimise treatment to surrounding healthy tissue (Alexander et al 1995). The treatment usually entails a single, high-dose fraction with very small planning and imaging margins. For these reasons, immobilisation is crucial and often involves a fixed frame attached to the patient's skull, or a comprehensive mask and imaging system. The use of surface-tracking for these patients has been investigated with a view to move away from the fixed-frame system allowing a noninvasive procedure for the patient, whilst other studies have investigated the use of an "open" mask system (Pan et al 2012, Li et al 2011, Wiant et al 2016). The use of the open-mask system has been evaluated well by patients in terms of anxiety and claustrophobia experienced during treatment (Wiant et al 2016). The improved patient experience alongside the possible reduction in radiation dose from imaging would show preference for this surface-tracking technique for position verification.

8.6.4.2 Limitations of Surface Tracking

Surface tracking has been discussed and reported to reduce the radiation dose received by a patient during their radiotherapy course, however, there are no studies suggesting a replacement of standard radiation imaging in favour of surface tracking. Following initial patient positioning, either using surface tracking or traditional skin markers, the patient's position must still be verified using standard on-treatment imaging, either 2-D or volumetric. Surface tracking cannot replace on-treatment imaging as there is no guarantee that internal organs will not move independently of external skin surface. Organ changes can occur specifically in lung patients (Kwint et al 2014) and in positioning of the prostate (Moreau et al 2017) during radiotherapy treatment. Surface-tracking equipment can be a useful tool to minimise the risk of irradiating healthy tissue due to patient movement when used alongside radiographic on-treatment imaging.

8.6.5 Implanted Transponders

As discussed for surface tracking, it is vital that a patient's position can either be maintained during treatment using strict immobilisation equipment or can be verified during treatment delivery by using equipment such as surface tracking or implanted transponders.

Immobilisation of the treatment or target area is difficult to achieve for certain areas of anatomy as external immobilisation will not restrict internal organ motion. This leads to uncertainty when the target volume is a mobile structure, such as the prostate. The external anatomy of the patient may not correspond to the internal location or movement of the prostate during treatment. On-treatment imaging using radiation, will not be suitable or achievable for continuous tracking of the prostate; using radiation will increase the radiation dose to the patient and surface-tracking techniques, although no radiation is used, cannot identify the position of the prostate itself. Implanted transponders do not require ionising radiation and are capable of continuous monitoring of the prostate (or the position of other target organs) throughout treatment delivery. The use of implanted transponders has concentrated on prostate radiotherapy; implanting the transponders in to the prostate is a relatively simple procedure and due to the prostate's mobility is beneficial to treatment delivery.

As the prostate is a moveable structure, CTV-PTV margins are important as the CTV is extended to the PTV primarily to account for geographical variations of the target (van Herk 2000). If the geographical uncertainty of the prostate can be minimised, the CTV-PTV margin can be reduced. This has been shown with the use of fiducial markers within the prostate (Chung et al 2004, Das et al 2013, Alander et al 2014). Although verifying the position of prostate prior to treatment delivery is seen as beneficial in the reduction of PTV margin, the margins cannot be reduced greatly due to the uncertainty of the prostate's position during treatment delivery. The use of transponders to monitor position, in real time, as treatment is delivered can give geometric information to inform a further reduction in PTV margins (Su et al 2011, Rajendran 2010, Chaurasia et al 2018).

Interoperator variability when using certain verification methods has caused limitations for some systems, ultrasound especially. The use of surgically implanted transponders can reduce this risk. The transponders are inserted prior to any radiotherapy planning scans being acquired. The position of the transponders is known during the planning stage and this position is used for daily verification and tracking of the organ. The position of the transponders is received by equipment software, there is little scope for interoperator variability, and, therefore, the information can be determined to be more reliable (Foster et al 2010).

The name of this equipment does highlight a limitation of the system: Implanted, the patient is required to have an invasive procedure to place the transponders in to the target organ or area before radiotherapy planning or treatment can begin. Any invasive procedure can provide complications and a patient may refuse the insertion of the transponders. Care should be taken to ensure any patient is fully aware of all the possible effects of having the transponders inserted as well as any possible detriment to the patient.

CHAPTER 9

Quality Systems and Quality Assurance

9.1 INTRODUCTION

In this chapter, we examine the principles and the framework for quality management systems (QMSs) and quality assurance (QA), as pertinent to on-treatment imaging for geometric verification. It is only the over-arching principles we will examine and their place in the practice of radiotherapy as a whole. More information for specific quality assurance, in the form of specific quality control checks, image quality, and dose, can be found in some of the published works referred to within this chapter and book. Here, the main themes are the framework principles for good, secure, and accurate on-treatment verification imaging for guiding radiotherapy.

9.2 DEFINITIONS—QUALITY MANAGEMENT SYSTEMS (QMSs)

The International Organization for Standardization (ISO) (DNV GL 2015) provides an international standard for quality management systems (ISO 9001:2015), developed with resources from individual countries in order to support and implement such standards, e.g. the British Standards Institution (BSI) (BSIGroup 2018). By definition, a QMS is a "set of policies, processes and procedures required for planning and execution (production/ development/service) in the core business area of an organization (i.e. areas that can impact the organization's ability to meet customer requirements.)" ISO 9001 is an example of a QMS (StandardsStores 2018).

In more straightforward terms, it is a set of interrelated or interacting elements that organisations use to formulate quality policies and objectives, and to establish the processes needed to ensure that policies are followed and objectives are achieved all through managing the activities an organisation might use to direct, control, and coordinate quality (Praxiom 2018). The system is the organisational structure, procedures, processes, and resources needed to implement quality management. There are seven key quality management principles involved, which are (in no particular order of importance):

- Customer focus

- Leadership

- Engagement of people

- Process approach

- Improvement

- Evidence-based decision making

- Relationship management

Although quite general in their outlook, we can see their relevance to precise and accurate radiotherapy and the nature of on-treatment imaging for ensuring that treatments are delivered geometrically as planned. For example, a process approach that would be systematic in its application, tailored for individual clinical sites; evidence-based practice ensuring the best and most appropriate protocols and equipment are used within available resources; and improvement in striving for the best and most appropriate image quality with minimised concomitant doses involved for the patient are relevant ways to apply three of the principles to improve organisational quality.

Within the UK, the health care sector has various requirements for needing quality management and/or an accredited quality management system in place. These include the CQC (CQC 2018), which continually examines all aspects of health care (from acute to social care), assessing an organisation's commitment to continuous quality improvement (CQC 2018). Within cancer services as a whole and radiotherapy in particular, an accredited QMS is required for complying with health care standards, now a part of the NHS's Quality Surveillance Programme (NHS 2018). The use of a QMS is recommended under a host of professional guidance documents within the UK and abroad (Podgorsak 2005; Hartman 2004; Emmerson et al. 2012; Hellebust et al. 2014). These include:

- In the UK, "Towards Safer Radiotherapy" (RCR 2008b)

- In France, "Management Guidelines for Safety and Quality in Radiotherapy" (ASN 2009)

- Within Europe, "Practical Guidelines for the Implementation of a Quality System in Radiotherapy" (Leer et al. 1998)

- Within the global community, the IAEA's "Accuracy Requirements and Uncertainties in Radiotherapy" (IAEA 2016b)

- And International Standards such as the International Commission for Radiation Protection standards ICRP 86 and ICRP 112 (Ortiz et al. 2000, 2009)

The latter two highlight how OMSs are recommended to help prevent major accidents in radiotherapy, including those involving newer technologies. Indeed within the UK, the use of a QMS in radiotherapy was piloted and initiated following one of these incidents. So the ultimate reason for engaging with a QMS for radiotherapy is that it is good for our patients.

Key aspects of the QMS include documents often denoted at various levels within the hierarchy of the system; as the levels increase, the subject matter moves from being more strategic to being more detailed regarding its part in the overall process. For example, the QMS will have:

Level 1 documents: The key over-arching quality policy declaring the intention of the organisation with respect to, for example: (a) clinical quality, (b) national standards and (c) national legislation; the scope of the system; the management structure and key roles and responsibilities within the organisation as a whole.

Level 2 documents: The main management procedures that discuss the clinical tasks required and the general actions needed to achieve the goals in the quality policy; they importantly also contain details of responsibilities for tasks at a more clinical level. Here there would likely be the main procedures for QC checks performed on on-treatment imaging equipment, and those for on-treatment imaging across a department.

Level 3 documents: The main detailed work instructions for how a task is undertaken, highlighting specific elements of a task (such as undertaking a protocol for on-treatment imaging in a particular clinical site; how the images are acquired, analysed, and acted upon; or how the quality control tests for, say, image quality or dose measurements are performed on the kV CBCT system), also indicating key documents and records that are required to safely perform the task, and record the results of such tasks.

As noted in the last point, records are vital to maintaining and managing the processes themselves, so results are known accurately and can be audited routinely to check their safety and validity.

All these documents are general guidance for ensuring that radiotherapy, throughout the world, is delivered as safely and accurately as possible; this is no less a requirement for on-treatment imaging as a major part of this process. But does this necessarily guarantee the quality of the clinical process? Not necessarily; the QMS can help with ensuring consistency of process within the system, but it is at the mercy of the integrity of the individual elements within the system, i.e. whether the individual protocols, processes, and procedures are clinically correct and effective themselves. If not, then the processes could be conducted incorrectly, but again and again in a consistent manner under the QMS, as has happened in the past with certain incidents. The quality of the tasks themselves is also paramount, which is where continuous audit, as part of the QMS, is vital to the whole essence of quality. However, it requires audit, which extends from ensuring that a procedure or work instruction has been followed accurately but also checks and audits the integrity and efficacy of the process itself for clinical quality. This is where clinical quality and quality assurance come into play.

Clinical quality is an aspect that is highlighted within the framework of clinical governance within the UK in the NHS, "a system through which NHS organizations are accountable for continuously improving the quality of their services and safeguarding high standards of care by creating an environment in which excellence in clinical care will flourish" (Scally and Donaldson 1998). It is enshrined in legislation in the UK (the Health and Social Care [Community Health and Standards] Act 2003, section 45), but is short on detail for specific aspects of health care. It does highlight that it is the duty of each NHS body, "to put and keep in place arrangements for the purpose of monitoring and improving the quality of health care provided by and for that body" (HSCA 2003).

There are seven "pillars" associated with NHS clinical governance and clinical quality. These are (a) service user, carer, and public involvement; (b) risk management; (c) clinical audit; (d) staffing and management; (e) education and training; (f) clinical effectiveness; and (g) clinical information. We can see some resonance with the principles of quality management, but here we are focused more on clinical quality and efficacy, service delivery, and management. These are key factors for the safe and effective treatment with radiotherapy, of which on-treatment imaging is a huge part, especially for today's complex technologies and techniques.

Within radiotherapy, professional guidance is regularly published, written by expert clinicians, physicists, and radiographers (radiation therapists), which can help guide the clinical work to ensure the highest quality standards are available to patients, however implementing such standards and techniques requires resources (equipment, staffing, training etc.) that are not always available because of cost implications and budgetary constraints in the UK. It also requires a good clinical evidence base in the forms of studies, publications and clinical trials. For the purpose of on-treatment imaging for geometric verification, professional guidance is available for both the clinical and technical assurance of quality. For example:

- UK guides for clinical practice (RCR 2008a; NRIG 2012)

- Reports for conducting commissioning and quality control checks (e.g. Bissonette et al. 2012; Kirby et al. 2006b; Brock et al. 2017; Patel et al. 2018; Ding et al. 2018)

- Recent Publications (e.g. D'Ambrosio et al. 2012; de Los Santos et al. 2013; Fontenot et al. 2014; IAEA 2016b; Zou et al. 2018)

Conducted under a programme of routine quality assurance, these guidance documents help clinical departments implement the best of current clinical and technical practice for on-treatment imaging for image guidance in radiotherapy; and indeed for all other aspects of the radiotherapy process itself. They are, however, only guidance documents; they are not mandatory, although performing quality assurance programmes is in many countries. But how is that achieved, and how is quality assurance defined?

9.3 DEFINITIONS QUALITY ASSURANCE (QA)

Quality assurance is defined in many different ways, often dependent upon the context in which it is applied. In general terms, the BSI defines it as "the part of quality management focused on providing confidence that quality requirements will be fulfilled" (BSIGroup 2018; DNV GL 2015). In the context of clinical health care in the UK, the Department of Health defined it as "a systematic process of verifying that a product or service is meeting specified requirements" (DoH 1990). This is still not very specific for our particular clinical context. The WHO in 1988 produced a definition specifically for radiotherapy, and the term QART (Quality Assurance in Radiotherapy) was defined as:

> Those procedures that ensure consistency of the medical prescription and the safe fulfilment of that prescription as regards dose to the target volume, together with minimal dose to normal tissue, minimal exposure of personnel and adequate patient monitoring aimed at determining the end result of treatment (WHO 1988; Leer et al. 1998; Emmerson et al. 2012).

This final definition gets to the heart of the matter for radiotherapy, and the essence of what one is trying to achieve through on-treatment imaging for geometric verification: To ensure that the prescribed dose and the associated dose distribution are delivered anatomically as prescribed so that the optimisation of prescribed dose to the target volumes and minimised (within defined constraints/tolerances) dose to the associated OARs is actually being achieved in vivo, for every fraction delivered through the treatment course.

Within the UK, quality assurance is required by law, under the IRR17 and IRMER 2017 legislative acts (statutory instruments). They are therefore a requirement of UK radiotherapy departments, but only as far as ensuring that a quality assurance programme exists (for the whole of the radiotherapy process, which includes on-treatment imaging) and is actioned routinely. There is no specific detail as to how it should be conducted; this is the premise of the professional guidance documents (listed previously in section 9.2). Within IRMER 2017, the definitions of quality assurance and quality control are set out specifically:

- "Quality assurance" means all those planned and systematic actions necessary to provide adequate assurance that a structure, system, component, or procedure will perform satisfactorily in compliance with generally applicable standards and quality control is a part of quality assurance.

- "Quality control" means the set of operations (programming, coordinating, implementing) intended to maintain or to improve quality and includes monitoring, evaluation, and maintenance at required levels of all characteristics of performance of equipment that can be defined, measured, and controlled.

One can often merely associate QA with only the checks performed on the radiotherapy equipment on a routine basis (daily, weekly, monthly checks etc.); but as we see in the definition, it goes far beyond this. QA includes these aspects and the clinical aspects, too; the processes of recording, audit, and review (as outlined under the QMS) are critical clinical aspects that are part of the QA of the on-treatment imaging process as a whole. Recognizing the importance of the technical checks on image quality, dose etc., but also the review of procedures and work instructions for clinically undertaking on-treatment imaging (image acquisition, frequency, analysis, set-up correction etc.) is an integral part of quality assurance. All of these aspects can be managed through the QMS to ensure the best possible practice and minimise the risk to the patients and staff. The art of quality audit, e.g. through the Deming Cycle (iSixSigma 2018), is a key component here, not just in auditing the quality system itself, but in terms of auditing the clinical processes and their effectiveness too. The cycle of "plan, do, check, and act" always completes with action to improve what was implemented; this is true for the process of the quality system, but can be applied to the clinical quality of the procedures and work instructions themselves. The goal is to improve clinical quality through, for example, the implementation of better, most appropriate on-treatment imaging techniques, improvements in image quality, more efficient and safe on-line imaging protocols, reductions in concomitant dose without reduction in clinical usability, and effectiveness for image matching etc.

9.4 DEFINITIONS—QA AND QMS PRACTICALITIES FOR ON-TREATMENT VERIFICATION IMAGING

For quality assurance pertinent to on-treatment imaging for geometric verification, one might consider there are two distinct areas:

1. Quality assurance for the technicalities of the equipment. These could include tests for image quality; concomitant dose for different imaging modalities and techniques; the accuracy of matching and registration algorithms, especially following software upgrades or new features; and the networking of images from, for example, TPS to OMS to linac and back again. As examples, see Mijnheer et al. (2004), Kirby and Glendinning (2006), Kirby et al. (2006), Bissonette et al. (2012), Morrow et al. (2012), Arumugam et al. (2013), Qi et al. (2013), Brock et al. (2017), Ding et al. (2018), and Samei et al. (2018). Additional details are given in chapter 13.

2. Quality assurance for the clinical use of the equipment. These could include review and audit of specific image acquisition and analysis processes; calculation and review of margins used for treatment planning (determined from an analysis of set-up errors for each clinical site); assessment of suitability of immobilisation equipment; qualitative reviews of image quality and repeat imaging, especially with respect to artefacts and noise on images; and reviews of overall concomitant dose and nonconformities regarding imaging frequency (specifically with respect to concomitant dose (RCR 2008a, NRIG 2012)).

Commissioning is the first process in the implementation of all new on-treatment imaging equipment, technologies, and techniques (see chapter 13). It forms the baseline data against which routine QA and QC checks are performed and establishes the protocols for clinical use and training etc. The documentation and rationale for how these operations are conducted are included in the QMS as part of the commissioning process.

Although written separately previously, the QA processes described are not mutually exclusive and require good recording, communication, and discussion around all these aspects in a multidisciplinary team dedicated to on-treatment imaging for image guidance (RCR 2008a, NRIG 2012). It is important that all staff (clinical and scientific) are involved in QA, recognise their part within it, are aware of system performance and tolerances for its optimal use, and for the best use of image guidance for the patient's radiotherapy (NRIG 2012). For example, routine assessment of imaging doses should be measured through routine QC checks (Patel et al. 2018), but then regularly reviewed and discussed, ensuring values or imaging frequency limits are reviewed for different clinical sites with clinical staff, especially the prescribing clinician (NRIG 2012). In this way, one ensures that the concomitant dose burden to the patient does not change significantly over time; and if it does, action is taken to reduce it (an action to improve quality of care). Image quality should be checked and assessed routinely in a similar way and results discussed more widely to ensure that in routine clinical use remains consistently good for the on-treatment imaging task of geometric verification (NRIG 2012). Again, if found to have changed significantly, action should be taken to remedy the image quality and/or reassess the clinical usage to perhaps use a different technology or technique. One of the most crucial aspects of assessing set-up error is its correction (on-line or off-line) to improve set-up. This depends upon image quality, but also upon the accuracy of the registration/matching software and its feedback to the linac for (automatic couch correction, for example). Quality assurance is needed to assess this accuracy and precision through QC checks (Kirby et al. 2006b; Arumugam et al. 2013; IAEA 2016b, Brock et al. 2017), and its implementation to couch adjustments for true and accurate geometric correction of the patient (Kirby et al. 2006b, Arumugam et al. 2013). This should be assessed at first commissioning, but also for software changes and upgrades. It is important, too, for clinical use and implementation, that clinical users are involved with accuracy of assessment and checks so that (a) one ensures that accuracy for different imaging modalities and techniques are consistent with reported data (that couch corrections take place and are appropriate in direction and magnitude) and (b) one ensures that training and competency are kept up-to-date for clinical users (NRIG 2012).

The detailed procedures listing the various tasks above and work instructions for how one achieves them would be documents contained in the QMS. Main on-treatment imaging tasks and those responsible for them would be part of the documented procedures (level 2 documents); whilst detailed information on how QC checks and clinical protocols are performed would be contained in level 3 work instructions. In all cases, regular audit would ensure that the procedures and work instructions are being followed as documented, and if not, would review and potentially correct, modify, or change clinical practice to bring processes back into alignment with documentation. The actual quality of the tasks could be reviewed, too, in light of QA results and review (above), watching for any excessive repeats in imaging task, changes in technology and techniques, internally and externally reported near-misses and clinical incidents, and allowing any evidence-based protocol reviews, etc. Here the benefit of regular audit, review, and potential change comes into its best work from the implementation of the QMS.

III

Advanced Issues, Techniques, and Practices

CHAPTER 10

Alternative Technologies

10.1 TOMOTHERAPY

10.1.1 Introduction

Having considered the most widely used forms of on-treatment imaging for geometric verification, i.e. by virtue that they are based on standard C-arm linear accelerators (linacs), we now start to explore the on-treatment imaging equipment and strategies on other types of advanced technology alternatively based on non-C-arm platforms. In this section, we begin to look at the TomoTherapy system and its on-treatment verification imaging capabilities. An excellent technical overview of the system is given in work by Langen et al. (2010).

10.1.2 System Overview

The helical tomotherapy system (TomoTherapy) was originally designed by a research group at the University of Wisconsin headed by Rock Mackie (Fenwick et al. 2006; Mackie 2006), and is now a commercial product of Accuray Inc., (Sunnyvale, California, USA) commercially available as the TomoTherapy® and Radixact™ systems. Following on from the ground-breaking development of the NOMOS "Peacock" system designed for a standard C-arm platform, which had an add-on binary MLC, dedicated inverse planned TPS, and specialised external couch drive mechanism (Fenwick et al. 2006; Mackie 2006), TomoTherapy engaged similar principles to design a single–unit machine based upon a CT gantry and technology. It was purposely built for image guided RT and IMRT, with its own binary MLC, dedicated inverse planned TPS, a CT (MV) detector subsystem, together with an in-line 6MV linear accelerator. Because it uses slip-ring CT technology, the couch can move continuously allowing for helical MVCT imaging (like a standard CT scanner) and helical treatment delivery, avoiding the slice-to-slice dosimetric junction errors associated with axial delivery (Fenwick et al. 2006, Langen et al. 2010). The historical development of the technology is well documented by Rock Mackie himself (Mackie 2006).

TomoTherapy® is a fully integrated system, with its own dedicated TPS, on-treatment imaging acquisition and registration software, as well as the ability for on-line rapid plan generation and adaptive treatments (Rong and Welsh 2011; Fenwick et al. 2006; Beavis 2004). The slip-ring gantry system allows full 360-degree continuous movement of the gantry, enabling treatment from any angle, although this is simplified in treatment plan optimisation to 51 equispaced treatment directions (approximately 7-degree spacing) (Rong and Welsh 2011; Fenwick et al. 2006; Beavis 2004). Beam generation is from a compact 40 cm long S-band 6 MV linac mounted vertically so the beam points constantly toward the gantry rotation axis at an SAD of 85 cm (Ouzidane et al. 2015). The system is flattening-filter free, so the maximum doserate is approximately 8 Gy/min at the isocentre (up to 10 Gy/min for the latest Radixact™ unit) (Ouzidane et al. 2015; Fenwick et al. 2006; Langen et al. 2010).

Intensity modulation at each beam projection is achieved by a 64-leaf binary MLC. Each leaf is individually and pneumatically controlled to rapidly open and close it; the x-ray beam is on when a leaf is open and off when closed. Each tungsten leaf is 10 cm thick and has a nominal width of 6.25 mm at the isocentre, providing a fan beam "length" laterally (x-direction, left-to-right) of 40 cm. In the y-direction (longitudinal), adjustable collimator jaws allow the beam "width" to be adjusted to either 1, 2.5, or 5 cm at the isocentre in the cranio-caudal direction (Ouzidane et al. 2015; Rong and Welsh 2011; Fenwick et al. 2006; Beavis 2004).

The gantry and the couch move at constant speeds during treatment; the gantry with a period of between 10 and 60 seconds per rotation, whilst the couch translates patients through the treatment beam at a constant fraction (pitch) of the fan-beam width (Rong and Welsh 2011; Fenwick et al. 2006). Helical pitch is therefore defined as the couch movement distance for a complete gantry rotation relative to the beam width at the isocentre, and typically lies between 0.2 and 0.5 (Fenwick et al. 2006) and allows for multiple intensity levels throughout the treatment volume. More complex dose distributions are possible at the expense of prolonged treatment times, the latter being determined by the overall target volume length and width, modulation factor, and dose per fraction. It is less dependent upon the pitch, which increases as the couch speed increases, but with a consequent decrease in gantry rotation speed (Rong and Welsh 2011). Nonetheless, these parameters must be carefully chosen to minimise rotational delivery effects such as the "thread effect"; a dose perturbation generated during helical delivery particularly at the periphery of the treatment volume (Takahashi et al. 2013; Fenwick et al. 2006). Dose delivery is based upon time rather than integral monitor units (Burnet et al. 2010); so intensity modulation can be achieved by leaf-specific opening times. During treatment planning, a modulation factor (MF) must be selected; this is defined as the longest leaf opening time divided by the average of all non-zero opening times. The longest opening time will determine the gantry rotation speed during treatment (Ouzidane et al. 2015).

Broadly speaking, an average TomoTherapy® procedure takes about 20 minutes of total time in the treatment room (patient entry to exit); this includes patient set-up, MVCT image acquisition, image registration and couch adjustment, and treatment delivery (Bates et al. 2013; Rong and Welsh 2011; Burnet et al. 2010; Thomas et al. 2010; Bijderkerke et al. 2008). These times can vary from about 12 minutes for relatively straightforward clinical sites like prostate, to about 25 for cranio-spinal treatments (Dean et al. 2013).

The longitudinal movement capability of the couch means a total cylindrical target volume of 160 cm (length) × 40 cm (wide) can be treated with intensity modulation under image-guidance on the TomoTherapy® system (Zeverino et al. 2012).

One of the latest capabilities of the TomoTherapy® system is (through its Radixact™ version) to offer both image-guided and nonimage-guided set-ups, for all intensity modulated and 3-D conformal treatments and rotational and non-rotational treatment options. One example of non-rotational treatments is commercially termed TomoDirect™. Here, instead of treatment being delivered helically, a discrete-angle, nonrotational option is offered primarily for 3-D conformal therapy. During treatment, all beams for each target volume are delivered sequentially with the couch passing through the bore of the unit at an appropriate speed for each gantry angle (Meyer et al. 2017; Han et al. 2016; Chung et al. 2015; Franco et al. 2014).

10.1.3 On-treatment Verification Imaging

In terms of on-treatment verification imaging, the technology is very similar to that of conventional helical CT scanners. The ring gantry contains the 6 MV linac x-ray source (reduced in energy to approximately a nominal peak of 3.5 MV, with a spectrum average of approximately 0.75 MV, Ouzidane et al. 2015; Burnet et al. 2010; Beavis 2004; Jeraj et al. 2004]), which is collimated to produce a fan beam of 0.5 cm (in the longitudinal y-direction) by 40 cm (in the lateral x-direction) (Chan et al. 2011). This type of arrangement is therefore called "fan-beam MV CT." Opposite to the linac is an arc-shaped array of 768 conventional xenon ion chamber CT detectors (Kinhikar et al. 2009b; Beavis 2004; Keller et al. 2002) at a source-to-detector distance of 142 cm with each detector projecting to a transverse width of 0.73 mm at the isocentre (Kinhikar et al. 2009a, 2009b). This arrangement has the advantage of acquiring volumetric on-treatment images postpatient set-up and prior to treatment delivery with a standard image matrix of 512 × 512 pixels per slice and FOV of 40 cm diameter (Ouzidane et al. 2015). There is a secondary benefit that lends itself well to adaptive radiation therapy—the detectors can collect data and back-project it through the acquired CT dataset to assess the actual dose distribution delivered to a patient during the fraction. The image quality (at a peak of 3.5 MV) is visibly poorer in contrast and spatial resolution with greater noise compared to that of conventional CT or CBCT scans taken at kV x-ray energies due to subject contrast differences between x-rays at kV and MV energies (Chan et al. 2011; Burnet et al. 2010); but the overall quality is acceptable for the purpose of on-treatment verification with reasonable image doses of approximately 0.4 to 3 cGy per volumetric scan (Bates et al. 2013; Rong and Welsh 2011; Burnet et al. 2010; Kinhikar et al. 2009b; Shah et al. 2008, Beavis 2004). This is highly comparable with MV EPID 2-D portal imaging (1 to 3 cGy per 2-D image) and kV CBCT 3-D volumetric imaging (3 to 5 cGy) (de Los Santo et al. 2013), making daily imaging feasible (Ouzidane et al. 2015; Beavis 2004).

The system is set-up to acquire a set of image slices for on-treatment verification imaging. By changing the pitch (the longitudinal distance that the couch moves during a single gantry rotation), these can have nominal slice thicknesses of 2, 4, or 6 mm, which are termed "fine," "normal," or "coarse" settings, respectively (Bates et al. 2013; Chan et al. 2011;

Rong and Welsh 2011). The acquisition rate is one slice per five seconds and reconstruction is concurrent with acquisition (Ouzidane et al. 2015; Bates et al. 2013). In many departmental protocols, the radiographers (RTTs) undertake and lead on-treatment verification imaging autonomously (often within a wider, multiprofessional team for both IGRT and IMRT/VMAT) (Bates et al. 2013; Dean et al. 2013; Burnet et al. 2010). This will include determining scan settings based on predefined protocols, with often a combination of settings being required over a treatment course; with the necessary close monitoring of associated concomitant radiation dose (Bates et al. 2013; Shah et al. 2012). The primary reason for this is that the spatial resolution of images is finer with smaller slice thicknesses, particularly in the longitudinal (couch movement) direction. Imaging dose increases and acquisition times lengthen with smaller slice thickness (Bates et al. 2013).

The time required for on-treatment imaging and analysis prior to actual treatment delivery is an important factor in busy radiotherapy departments; contributing to the overall time slot which is allocated to each patient (Bates et al. 2013; Burnet et al. 2010; Thomas et al. 2010; Bijderkerke et al. 2008). Acquisition (or scan) time can be regarded as the time between the radiographers (RTTs) preparing to acquire a scan and the completion of processing ready for registration. Registration times are then from the point of completion of image acquisition to accepting any necessary couch shifts, prior to beam on (Bates et al. 2013; Bijdekerke et al. 2008). The image matching procedure (Bates et al. 2013; Burnet et al. 2010) makes use of preagreed match structures and anatomy, as suggested in earlier forms of 2-D image matching (RCR 2008a), and depends upon contrast between fat, soft-tissue, and bony structures within the primary matching site. Within the TomoTherapy system, rapid automatic matching (using automated bone and soft-tissue protocols) are often initially used followed by manual refinement for optimising the match. Action levels for positional correction can be 1 mm for translations and 1 degree for rotations (Bates et al. 2013).

Within Bates et al.'s (2013) excellent in-depth study of imaging protocols for TomoTherapy, they found that practice gradually changed from using predominantly "fine" (2 mm) scans (with a consequently higher concomitant dose) to using mostly "course" (6 mm) scans, over a five-year period of experience. Overall scan acquisition time is highly dependent upon the length of the scanned volume. For average clinical sites of prostate, pelvis, head and neck, and CNS, the scan lengths were 5.9, 12.2, 10.3 and 5.9 cm respectively with scan acquisition and matching times ranging from 3.8 to 8.6 min., 5.9 to 13.4 min., 5.3 to 12.0 min., and 3.8 to 8.6 min., respectively (the range being across the scanning mode of "coarse" to "fine"). Studies are on-going to examine increasing on-treatment verification times by increasing scan length in certain sites and moving from "coarse" to "normal" slice thickness for others with estimated increases in the overall treatment day of two to three minutes (Bates et al. 2013).

Commissioning and implementation are very important considerations for a technology like TomoTherapy that is so different in design and clinical application to standard C-arm linacs, for which there is a wealth of information and experience available. Perhaps one of the most comprehensive reports on the overall preparation and implementation of a clinical TomoTherapy service can be found in the research by Burnet et al. (2010); a paper which, in its foundational principles, can be applied to the commissioning and implementation of any new piece of technology. At the time of writing, the set-up and training for on-treatment imaging, which is a necessary part of the clinical use of TomoTherapy, was new for the whole department; now one would hope that with the widespread use of on-treatment verification imaging, implementation of that aspect of TomoTherapy should seem slightly less novel.

Burnet et al. (2010) outline several key areas of focus for commissioning and implementation of TomoTherapy, which include:

- Commissioning and quality assurance/control
- Staff (all) education and training
- Protocols and treatment pathways
- Treatment planning system commissioning
- Daily on-treatment MVCT verification
- Patient selection and ramp-up of daily workload
- Back-up treatment plan strategy
- Use of phantoms and *in-silico* case studies

Regarding on-treatment verification imaging, key highlights include training for all professionals, use of phantoms for establishing image acquisition and matching protocols and developing experience, multiprofessional team involvement in concomitant dose issues (and discussion of risk with all patients during the consent process), recording of imaging protocols (particularly with regard to dose factors) in a database, establishing action levels for patient set-up (Burnet et al. 2010), patient selection and ramp-up of patient workload and overall treatment time analysis (Dean et al. 2013; Burnet et al. 2010; Bijderkerke et al. 2008).

As detailed briefly above, MVCT scan acquisition is dependent upon the chosen scan length and scan settings (nominal equivalent slice thickness), which will develop over time depending upon experience of image matching, dose and overall treatment time considerations (Bates et al. 2013; Dean et al. 2013; Burnet et al. 2010; Bijderkerke et al. 2008). The image-matching process is a two-stage one; first using a rapid automated match (utilising bone and soft-tissue protocols assessing six degrees of freedom), which establishes any gross error detection with action levels of between 2 and 15 mm for translations and 2 to 5 degrees for rotations (Burnet et al. 2010). A gross error detection would trigger a reset-up of the patient and a rescan – but this is a rare occurrence of only in approximately. 0.3% of patient scans (Bates et al. 2013; Burnet et al. 2010). After the initial match, the process is repeated automatically (assessing four degrees of freedom) and with final manual refinement, with action levels of 0 to 1 mm for translations and 0 to 1 degree for roll (Bates et al. 2013; Burnet et al. 2010). Initial results can identify substantial random displacements compared with the CT planning scan for most patients, including those within thermoplastic immobilisation; these are corrected on-line before treatment delivery. Image quality, speed of acquisition, and matching are all refined as greater experience is gained with the technology, often resulting in a gradual ramp–up of patients and reduction of overall treatment time (Bates et al. 2013; Dean et al. 2013; Burnet et al. 2010; Thomas et al. 2010; Bijderkerke et al. 2008). In some centres, the ramp-up has been much more rapid, dependent upon the workload case-mix. For example, those centres using TomoTherapy for predominantly prostate patients have shown a rapid rise in workload (to approximately 36 patients over an eight-hour working day) (Dean et al. 2013), where the scan and acquisition times can be three to four minutes (Bates et al. 2013) and the delivery quality assurance (DQA) conducted for each patient can be very quick (10 minutes after phantom set-up) (Dean et al. 2013).

10.1.4 Some Clinical Perspectives

On-treatment verification imaging is an integral part of the TomoTherapy system (Burnet et al. 2010; Zou et al. 2018) and therefore is a standard requirement for nearly all treatments, irrespective of complexity. It therefore lends itself to being able to monitor tumour responses daily and assess organ motion/changes interfractionally throughout the treatment course (De Los Santos et al. 2013). But one must remain acutely aware of the concomitant dose burden to the patient involved in this. Clinically, it has been used for numerous sites (Bauman et al. 2007; Burnet et al. 2010; De Los Santos et al. 2013) including head and neck (Chatterjee et al. 2011); prostate (Fiorino et al. 2008); breast (Franco et al. 2014; Haciislamoglu et al. 2015; Kaidar-Person et al. 2016); and cranio-spinal and total body/marrow techniques (Kim et al. 2017a; Yoon et al. 2011; Zeverino et al. 2012; Magome et al. 2016; Schultheiss et al. 2007).

The imaging is fully integrated into the system (Fenwick et al. 2006) and so with imaging at MV beam energy, the accompanying treatment-planning system can take account of and calculate the associated concomitant imaging dose. Since on-treatment imaging is with the actual treatment beam reconstruction of the delivered dose in vivo is possible (De Los Santos et al. 2013), allowing measurement of the actual delivered dose to the target volumes and nearby OARs. Studies of this nature have been conducted for clinical sites like prostate (Kupelian et al. 2006) and head and neck (Han et al. 2008; Lee et al. 2008)—sites where anatomical changes can readily occur, either interfractionally (head and neck) or intrafractionally (prostate). This forms the ideal basis for adaptive radiotherapy—where daily changes can be quantified for their effect on the delivered dose distribution (actual rather than computed/simulated). Coupled with deformable image registration algorithms and modelling, treatment plans can be adapted to account for dose delivered to different parts of the target volume (Chen et al. 2006; Lu et al. 2008).

Although, as noted above and in chapter 7, imaging with an MV energy beam means that there is likely a higher concomitant dose associated with on-treatment verification imaging, a poorer subject contrast, especially for soft tissues (compared with kV energy imaging), but fewer imaging artefacts in the presence of high Z materials (like hip prostheses and dental fillings). An advantage of its integration into the treatment delivery unit is that on-treatment verification imaging can be easily used at the beginning, in the middle, or at the end of the given fraction. The need for rigorous quality assurance for treatment delivery, geometry and dosimetry, and imaging quality/concomitant dose are particularly of note (Mege et al. 2016; Moore et al. 2010; Westerley et al. 2012; Chen et al. 2013; Fenwick et al. 2004; Nobah et al. 2014; Sen and West 2009; Balog and Soisson 2008; Bissonette et al. 2012; Jung et al. 2012).

10.2 CYBERKNIFE

10.2.1 The Cyberknife System

The Accuray Cyberknife system (Accuray Inc., Sunnyvale, California, USA; www.accuray.com) is a treatment technology based upon a robotic delivery system, which is used widely for industrial applications, such as the car and automotive industry. The whole treatment unit consists of distinct subsystems, based around a lightweight compact X-band 6 MV linac, a linac robotic manipulator, a similar robotic couch and a cross-fire (orthogonal) 2-D kV on-treatment imaging system, similar to that used in ExacTrac (see chapters 7 and 8). The combination of in-room kV imaging and the highly versatile and precise robotic manipulator (which has full six degrees of freedom and a 0.12 mm precision for the present system) means that it is primarily designed for frameless stereotactic radiosurgery—both intracranial and extracranial. It allows for a very high degree of accuracy and precision for the treatment geometry, but requires sophisticated on-treatment imaging techniques using both ionising and nonionising radiation to ensure accuracy of clinical delivery, even though the mechanical movements of the robotic manipulators are reproducible to within about 0.12 mm (Hara et al. 2007; Antypas and Pantelis 2008; Kilby et al. 2010).

The X-band linac runs flattening filter free (FFF) with an in-line, compact, sealed gun-standing waveguide tungsten target system delivering approximately 1000 MU/min. The 160 kg linac subsystem also has a sealed air-filled ionisation chamber, primary collimator, and a developing set of secondary collimation systems. These are at present:

- Twelve separate, fixed, circular collimators ranging in diameter from 0.5 to 6.0 cm.

- An IRIS variable collimator.

- A high definition (0.25 cm) Multileaf Collimator (M6 INCISE).

Because of the accuracy of the robotic systems, the 12 fixed collimators and the IRIS variable collimator can be stored on a specifically designed table fixed in a precise location within the treatment room. The system software is programmed with the room dimensions and coordinate system, so it can interchange the collimators entirely by itself during treatment delivery; thus optimising dosimetry and treatment time. The IRIS collimator can produce variable collimation from a 'closed' setting of .025 cm to a fully open 6.8 cm diameter field; but in practice, settings equivalent to the fixed collimators are used for planning and delivery (Hara et al. 2007; Kilby et al. 2010; Pantelis et al. 2012). The INCISE M6 MLC is the newest secondary collimation system offered, consisting of 82 tungsten leaves (41 leaf pairs) covering a field size of 10 × 12 cm and leaf width of 0.25 cm at 80 cm

SAD (Fahimian et al. 2013; Jang et al. 2016; Kathriarachchi et al. 2016; Kim et al. 2017c). The manufacturer quotes a positioning accuracy of within 0.5 mm and a reproducibility of 0.2 mm. It thus provides the flexibility of high-definition, irregular shaped fields with the accuracy and six degrees of freedom of robotic manipulation to deliver non-isocentric and non-coplanar beams.

The linac and couch robotic manipulators are highly accurate, with a repeatability specification which has improved over the years from 0.5 mm to now within 0.12 mm – a challenge for present linac based delivery and patient support systems. The movements of both manipulators are under computer control. The linac manipulator is configured to direct radiation to the volume of space around the beam intersection of the kV imaging system; the robotic couch positions the patient accurately within this region, accessible by the radiation beam. Fine precise movements are possible, and set-up correction is achieved by movement of the linac and/or the couch system. Under continuous or near continuous x-ray on-treatment imaging guidance, this means that there is the potential to track target movement and/or make intrafractional adjustments during the course of the treatment fraction.

Since the linacs movements are not constrained by a gantry, then most treatments can and indeed are, non-isocentric. There is a geometric reference point within the treatment room, like a traditional C-arm linac, but the 6 MV treatment beam is not physically constrained to always point towards it. The central reference point serves as the physical and mathematical origin for several coordinate systems used by Cyberknife, to which the robotic manipulators and the imaging system is calibrated and defined. For calibration, this point in the room is defined by an "isocrystal" (a light sensitive detector) mechanically mounted onto a rigid post called the "isopost". This point is space is defined as the 'geometric isocentre', but is not the same as the treatment isocentre and the treatment beam is not confined to always be directed towards it (Antypas and Pantelis 2008).

This means that there is a tremendous degree of flexibility possible in terms of beam directions, their size (using the circular collimators) or shape (using the MLC), their intensity (based upon the exposure time) etc. (Hara et al. 2007; Antypas and Pantelis 2008; Kilby et al. 2010). True dose painting is possible and indeed necessary to cover target volumes with a uniform or intentionally non-uniform dose; even those with concavities, intricate shapes and in close-proximity to OARs. Treatment Planning makes use of all conventional imaging modalities (CT, PET-CT, 4D CT, MR) within a dedicated treatment planning system (Multiplan). Standard image fusion and outlining is used for defining PTV and OARs and inverse planning is performed through Monte Carlo optimisation. The TPS begins with a pre-defined initial set of beam configurations based upon the target volume geometry. Each beam is a vector based upon a source point (the x-ray source) and a direction point (within the target volume). Each source point is described as a *node* and the complete set of nodes is called a *pathset*. There are therefore an infinite variety of nodes and pathsets which might be used to direction radiation onto the target, although there are naturally some mechanical constraints within the system and of course clinical dose constraints requested for the plan. The system will therefore try to optimise the various elements of nodes, pathsets, fieldsizes and weights, intensity modulation etc. in order to satisfy the dose planning constraints. A further compromise is required, in terms of optimisation of plan against delivery time – so intensity modulation (by multiple beams delivering to unique points in the target volume) and plan quality compared with movement during irradiation through the pathset. For the latter, a further optimisation (path traversal) algorithm is used, to sequence the nodes through the pathset for the speediest delivery. The optimisation engine will generally start with thousands of possible beams in a pathset; a typical treatment will consist of approximately 100 different beams per fraction.

10.2.2 On-treatment Verification Imaging Systems

In one way, because the Cyberknife system is non-isocentric with a large degree of flexibility in terms of position and direction of treatment beams, on-treatment imaging is vital to ensure the highly precise positioning capability of the system is then delivered geometrically accurately to the patient. The primary imaging technology used for this is in-room 2-D kV energy x-ray systems.

Two diagnostic quality x-ray tubes are mounted in the ceiling of the room, whose beam directions are orthogonal to one another and at 45 to the floor. Mounted on the floor are two corresponding amorphous silicon flat panel imagers set-up either at 90 degrees to each imaging beam or set within the floor, flush to the surface (so at 45 degrees to the imaging beam directions). Typically the x-ray source to detector distance is about 3.5 m. The beams form a cross-fire arrangement, so that the Cyberknife targeting and the imaging alignment centre is defined by the isopost – the centre of the coordinate system reference for the Cyberknife system, the geometric isocentre. The patient is ideally placed with their clinical target volume at or close (within about 10 cm) to the intersection of the imaging beams and hence the geometric isocentre; but the treatment isocentre does not need to be at this point – the treatment plan and associated reference images (DRRs) relate the treatment isocentre to the geometric centre and the image alignment centre (allowing for positional tolerances). It relies heavily on secure, rigid geometry between the geometric centre and the image alignment centre for treatments to be delivered accurately and precisely as planned. This requires a routine and robust system of commissioning and on-going QA (see chapters 13 and 9 respectively). So mounting of the robotic systems and the imaging components (x-ray tubes and imaging panels) need to be robust and secure; and maintained as such over time. For the current system, the flat panel imagers are approximately 41 × 41 cm square, and mounted flush in the floor. Because of this, software corrections are required (in real-time) to remove the distortion from the images, since they are not perpendicular to the beam direction.

In addition to x-ray imaging, nonionising radiation methods are used, often in consort with x-ray imaging, especially to exploit more fully the tracking capabilities of the Cyberknife system where there is likely to be significant intrafractional motion – i.e. in clinical sites involving the lungs and respiration. An optical tracking system is also available, which uses a stereo camera system which can track optical markers placed on the patient's skin or surface during treatment. These are often used in conjunction with the x-ray on-treatment imaging system in, e.g., their Synchrony Respiratory Tracking System (see section 10.2.3). Once more, positional accuracy is key, through careful calibration and continuous QA, to ensure that the optical images of the markers can be related to the geometric isocentre of the room and the imaging alignment centre for the x-ray images.

10.2.3 On-treatment Imaging Capabilities

The main reference images used for initial and beam alignment are a pair of 2-D DRRs; reconstructed for the directions of the cross-fire kV imaging beams in the room. These are obtained from the CT planning scans during pretreatment planning; so like most C-arm gantry based 2-D imaging procedures, the reference views are orthogonal to each other, but unlike the C-arm systems, they are non-orthogonal to the patient. The views will be similar to the principles described for Exactrac in chapters 7 and 8. Beam alignment at the start of treatment is based on automatic registration of the DRRs with live images acquired in the treatment room. From these 2-D imaging pairs, 3-D transformations are computed so that the patient can be translated into a correct set-up – since the geometric relationship between the in-room x-ray imaging system and the geometric isocentre is known, and so too (from the treatment plan) the expected relationship between the anatomy around the target volume and the intended treatment delivery plan (the treatment isocentre). These initial adjustments are made using the robotic couch at the start of treatment.

Further 6D corrections are possible throughout the delivery of treatment intrafraction-ally. As with any feedback correctional system, there are ideal tolerances and limitations to the extent of geometric correction in real-time, in order to maintain positional accuracy. For the Cyberknife system these are up to about +/− 2.5 cm for translations and about +/− 5 degrees for rotations – so initial set-up is used to ensure that any likely intrafractional corrections needed will be within these tolerances. During treatment, further additional translations and rotational corrections are achieved by feedback to the robotic manipulator of the linac to compensate for small target movement or, in the case of target volumes which move due to respiration and other internal organ motion, to track and follow the position during treatment. Typically 2-D kV image pairs can be acquired, the target localised and alignment corrections made every 30–60 secs; but clearly with a close monitoring of the overall concomitant dose burden involved for the patient.

In terms of on-treatment image guidance, several methods and algorithms are employed to register and track anatomy for initial set-up and during treatment. These are continually being refined and developed, but currently include:

- Bony anatomy registration and tracking. E.g. skull tracking (for intracranial SRS with full 6D movement and head and neck targets which may be considered rigidly related to the skull) where the high radiographic contrast of bony features with kV energy x-rays is utilised (Kilby et al. 2010; Fu and Kuduvalli 2008).

- Spine Tracking. Used for targets which within or near the spine, which use the high radiographic contrast of the bony detail as its main characteristic (Fu and Kuduvalli 2006; Fu et al. 2006; Ho et al. 2007; Muacevic et al. 2009; Furweger et al. 2010; Kilby et al. 2010; Furweger et al. 2011). The spine can be analysed and tracked either as a rigid body or as a non-rigid one, since the vertebrae can naturally move independently of one another.

- Fiducial markers. E.g. using implanted fiducials (usually 1 to 5 cylindrical gold seed markers (approximately 0.8 to 1.2 mm diameter, 3 to 6 mm length) are implanted into soft tissue for initial set-up and real-time tracking for soft tissue targets such as lung, liver, prostate and pancreas (Reichner et al. 2005; Mu et al. 2006; Anantham et al. 2007; Hatipoglu et al. 2007; Kilby et al. 2010; Ouzidane et al. 2015; Chan et al. 2017). The position of the fiducials are identified within the DRRs from the planning CT scans, and registered automatically with the markers' locations detected in the in-room stereoscopic on–treatment images.

- Lung tracking. Possible using implanted fiducials as above (Reichner et al. 2005; Anantham et al. 2007) or directly of the lung tumour itself depending upon its size, location and overall density compared with the surrounding lung tissue. Localisation is a two stage process using a global find for the target volume at initial set-up followed by loco-regional searches around the lung tumour during treatment delivery. Accurate identification and tracking appears possible in about 50% of cases, usually for those tumours larger than about 15 mm diameter and depending also upon whether the on-treatment imaging views are obstructed by more dense tissues and organs (e.g. heart and spine) (Kilby et al. 2010; Ouzidane et al. 2015).

- Synchrony Respiratory Tracking. Methods which are used for target volumes with expected movement (e.g. lung tumours or those affected by diaphragmatic motion). For synchrony, the main concept is to combine and correlate the positions of light emitting external markers placed on the patient's chest viewed by the stereoscopic optical tracking system. The positions are correlated with dynamic stereoscopic 2-D x-ray images acquired on-treatment using linear or non-linear modelling of the motion. This combination of optical external marker imaging and internal x-ray anatomical imaging enables a motion model to be developed, which then is feedback to Cyberknife to dynamically move and position the linac and treatment beams during breathing motion. (Shirato et al. 2004; Keall et al. 2006; Sayeh et al. 2007; Seppenwoolde et al. 2007; Kilby et al. 2010; Ouzidane et al. 2015). The correlation model is continually updated throughout the fraction dependent upon the motion detected in the optical and x-ray on-treatment images.

- InTempo Adaptive Imaging. Used for clinical sites where the imaging frequency needs to be adapted because of potential unexpected movements during the fraction, e.g. the prostate. The concept is to change the frequency of x-ray images dependent upon the anticipated and/or detected motion within the live images; the frequency of imaging and re-alignment is therefore changes in response to the rapidity of target motion. During shallow or slow drift, the imaging frequency can be slow. Once rapid (amplitude) movement is detected, the imaging frequency is increased to monitor and correct for the displacement. Once stable positioning is regained, the imaging frequency is then automatically reduced accordingly. In this way, on-treatment concomitant imaging dose is reduced.

With this array of target and motion tracking methods, the overall accuracy of localisation and correction of set-up error may be compromised. Independent studies have shown that targeting accuracy is submillimeter in most cases, thereby allowing margins to be reduced around the CTV. For those sites not significantly affected by respiratory movement, accuracy can be of approximately 0.5 mm has been measured in phantoms and clinically for the variety of skull, spine and fiducial marker techniques (Kilby et al. 2010; Antypas and Pantelis 2008; Ho et al. 2007). Those involving respiratory motion, phantom and clinical studies describe an accuracy achievable <1.5 mm (Dieterich et al. 2011; Kilby et al. 2010).

Treatment delivery times are very much case dependent and also depend upon the on-treatment imaging and tracking techniques used. Clinical sites have included prostate/pelvis, lung/thorax, gynecological, brain, spine and SBRT/SRS procedures (De Los Santos et al. 2013). They are typically about 30 min for target volumes not affected by respiration and about 45 min for those utilising synchrony and other tracking techniques; but this is highly dependent upon clinical site, target volume location and complexity of treatment plan delivery (Kilby et al. 2010; Ouzidane et al. 2015). One would anticipate that continued advancement would be through improved image detectors and image processing/analysis algorithms (Dieterich et al. 2011) and through new techniques such as dual kV energy imaging and image subtraction (Hoggarth et al. 2013) which could improve soft tissue visualisation (Hoggarth et al. 2013, Dieterich et al. 2016a). More efficient detectors would likely contribute to more accurate automatic detection and matching routines, with an associated reduced concomitant dose burden.

10.3 GAMMA KNIFE

10.3.1 The Gamma Knife System—Introduction

Perhaps the most traditional dedicated equipment for delivering stereotactic radiosurgery, particularly for intracranial lesions, has come over many years in the form of the Leksell Gamma Knife (LGK). It has developed from a single device in Sweden to being marketed globally, having successfully revolutionised neurosurgical practice from something that by necessity was always invasive with all the associated clinical risks of craniotomies and open-skull surgery. The original principles of stereotaxy and surgically pinned headframes (which are still very much used today), were first developed and published by Swedish neurosurgeon Lars Leksell in 1951. Developed at first for highly accurate and precise neurosurgical techniques, he then pioneered the use of radiation (using kV x-rays, protons, and eventually Co-60 gamma ray photons) for performing intracranial surgical procedures noninvasively, using high doses of radiation (Goetsch 2015a, 2015b).

Stereotactic radiosurgery began exclusively for cranial lesions, but is now applied extracranially too. For cranial work, procedures include single-fractionation using pinned headframes; single and multiple fractions using relocatable headframes and also now frameless processes too. The precision possible with pinned headframes (Warrington 2006) is now approached for some frameless systems (ExacTrac) using on-treatment image guidance. The use of Cobalt sources for the dedicated unit (hence the name Gamma Knife) began clinically in about 1967 in Sweden, with some of the first designs shipped to Buenos Aires,

Argentina and Sheffield, UK in the early 1980s (Goetsch 2015b). The newly designed 201 Co-60 source LGK was installed in UPMC in the late 1980s. The basic principle was that through a series of collimator helmets (which had 201 extremely well-machined collimator holes), the individual sources formed beams pointing toward a common focus. The patient, with their heard pinned within the stereotactic frame, was moved very accurately to different positions so that the radiation focus effectively produced individual "shots" of dose, so dose was painted to different parts of the target volume within the brain (Goetsch 2015a, 2015b). The near spherical dose "shots" were produced by the different collimator helmets, which had fixed diameters for all of the sources of 4, 8, 14 or 18 mm. Treatment positioning and verification was very much a manual task, with the operator making the fine translational movements, verified by a second person, and operators swapping over the different collimator helmets (if required) to deliver the treatment plan. So verification was entirely manual with no automatic or interlocked procedures; errors of set-up could remain errors of set-up without detection (Goetsch 2015b).

10.3.2 On-treatment Verification—First Automation

A key development was the use of an automatic positioning system (APS) in 1999, which allowed the patient's head to be moved with high-precision motors and encoded verification of the movement. This enabled multiple shots without user intervention and user verification, sped up treatment planning and delivery, and allowed more conformal treatments (which could be delivered in a sensible timescale) (Kuo et al. 2004; Goetsch 2015b). In one study (Goetsch 2002), 15 self-reported treatment errors on the previous system (wrong collimator, wrong coordinates, shots treated twice) could potentially have been avoided following the introduction of the APS and new verification system. Around a similar time, the first clinical experience was being gained with multiple fraction treatments (whilst still using pinned frames) on the LGK (Goetsch 2015b) although relocatable frames were being developed for SRS and SRT applications on a standard C-arm linac (Kassaee et al. 2003). On-treatment verification here was still a manual task, relying on skilled operators and relocation of pinned frames with respect to the skull for multiple fractions. No on-treatment verification imaging was possible to verify and correct for any movement with respect to the original treatment plan.

10.3.3 The Gamma Knife Perfexion (GKP) System—Further Automated Verification

In 2006/7, the Gamma Knife Perfexion (GKP) was first introduced, with many describing it as a paradigm shift from the original version. For this version, 192 Co-60 sources were used, which were able to move on the outside of a single stationary collimating structure that still provided collimated beams of 4, 8, and 16 mm diameters (Goetsch 2015b; Cevik et al. 2018). No collimator helmet changeover was required, and the sources were mounted on eight sector plates, which could be moved independently of one another automatically; each sector could produce 24 cobalt beams of different diameters (instead of all the sources being restricted to the same collimation diameter at any one time). Since each sector can move, positioning the sources in-between collimation holes effectively turns the beams "off," thus reducing dramatically the radiation protection issues for patients and operators.

The volume of space within the collimating structure was increased considerably, so that extracranial targets (head and neck) could now be treated, especially when coupled with a new APS (now termed the patient positioning system, or PPS) with automatic verified movement control that was submillimeter (better than 0.2 mm, with reproducibility better than 0.05 mm). PPS and source-sector movements are now fully automated and mechanically verified with dynamic control of PPS and sources during treatment delivery. This has meant that treatment times have been reduced from about 80 minutes to 20 minutes; set-up time reduced from 10 minutes to about two minutes; and collimator set-up time reduced from about 7 to 10 minutes to less than three seconds (Elekta 2015; Goetsch 2015b; Elekta 2018c).

Even with these developments, most procedures are still single fractions, since the patient is in a pinned headframe and on-treatment verification relies upon mechanical precision of the movements and encoded engineered readback. It also still makes the broad assumption that the internal, soft-tissue targets are in a fixed, rigid, geometrical relationship with respect to the skull, and that the skull is rigidly attached to the headframe, so that stereotactic principles relating to the frame (and hence the skull) would therefore relate to localising the target within the brain. Relocatable frames have been developed for SRT fractionated SRS, but with mixed results for accuracy between relocatable and pinned frames (Ruschin et al. 2013). Additionally, for extracranial target volumes in the head and neck and the extended reach of the GKP, using a relocatable bite-block designed frame (Ruschin et al. 2010), internal anatomical changes are more likely, which still cannot be verified and corrected for geometrically (to maintain the accuracy achieved by rigid frames) without on-treatment imaging of some description.

10.3.4 On-treatment Imaging—Development of CBCT

With the advent of 3-D kV CBCT and in-room kV stereoscopic imaging on C-arm linacs (see chapters 7 and 8), linac based SRS and SRT was in a position to improve frameless procedures to a geometric accuracy approaching that of pinned frames. In 2013, Ruschin et al. (2013) proposed an on-treatment imaging adaptation for the GKP system by mounting a kV x-ray tube and flat-panel imager onto the front face of the system to acquire 3-D kV CBCT images for on-treatment geometric verification. Mechanical integrity and image quality was found to be excellent, with the system being used for both single and multiple fraction treatments; maximum targeting errors were estimated to be about 0.4 mm. The system has been developed into a full commercial add-on for the GKP called Icon (Elekta 2018c). Attached to the unit is a specialised (combined) imaging arm with combined x-ray tube and imaging panel, which can be deployed into position and rotated around the patient (Elekta 2015). Since imaging isn't possible through the PPS, the rotational scan range is limited to about 200 degrees, but image quality is still assessed as acceptable for geometric verification and to visualise internal soft-tissue structures, especially for fractionated treatments (Reiner et al. 2017). The scan time is about 30 seconds and the imaging arm (which is articulated to move it into position and execute a scan in a unique fashion) has a source to detector distance of 100 cm (White Paper 2015, Wright et al. 2017). Typical clinical procedures and protocols are similar to CBCT for C-arm linacs, with possibilities for both inter- and intrafractional set-up error analysis when using CBCT before and after the treatment fraction, thus evaluating residual errors (Li et al. 2016). Impressive submillimeter (mean <0.2 mm) results have been obtained for the differences between pre- and post-treatment scans for intracranial SRS patients (Li et al. 2016).

In addition to CBCT x-ray imaging, the commercial Icon system employs a stereoscopic, optical on-treatment imaging technology termed the high-definition motion management system, or HDMM (Elekta 2015; Elekta 2018c; Stieler et al. 2016; Wright et al. 2017), particularly for improving the assured accuracy of frameless treatments. This is a stereoscopic infrared camera system used to continuously monitor the patient during irradiation by means of reflective markers on the patient's nose relative to four fixed markers on the GKP head-support system. Any movement (beyond a user defined threshold) detected by the HDMM automatically interrupts irradiation. The recent study by Wright et al. (2017) showed that the nose tip was a reasonable surrogate for cranial movement during treatment for frameless treatments, with no patient within the study (n=11) expressing intracranial target displacements (as assessed by CBCT) exceeding nose tip displacements (Wright et al. 2017). Other works have also verified the potential improvement in set-up accuracy for fractionated Gamma Knife treatments using the on-treatment imaging capabilities of Icon (Stieler et al. 2016; Reiner et al. 2017); even to the extent of on-line adaptive replanning following CBCT imaging (Stieler et al. 2016).

With the improved on-treatment verification imaging capability and the treatment delivery innovations of the GKP system, there are continuing indications (single and multiple brain metastases [Greto et al. 2016]) with maintained or improved QOL [Miller et al. 2016; Bragstad et al. 2018]) and emerging indications for fractionated GK RS procedures that extends the clinical compass to treating, for example, large meningiomas, perioptic tumours, and vestibular schwanomas (McTyre et al. 2017). Treatment-plan quality has been compared with modern linac-based VMAT techniques using FFF treatment beams (Abacioglu et al. 2014), with good results for target coverage, homogeneity, and OAR sparing, but with considerably different treatment times (nearly a factor of 10 times longer for the GK treatments). Treatments using GK, whether framed or frameless, still remain effective and well tolerated (Jacobs et al. 2018).

10.4 HALCYON

10.4.1 The Halcyon System—Introduction

Halcyon is perhaps the newest of the major treatment delivery platforms commercially available. Launched in 2017, its technology is based on an O-ring design, like TomoTherapy and Vero (Cozzi et al. 2018; Li et al. 2018; Michiels et al. 2018). It is designed to use the enhanced speed of movement capable with current technologies, and also work around legal limitations of gantry rotation speed (1 rpm) for open C-arm linac platforms for treating patients. Like TomoTherapy and Vero, all moving parts on the gantry are maintained behind closed covers so the faster rotation speeds possible for the O-ring designs can be exploited for both treatment delivery and for on-treatment imaging (Michiels et al. 2018). The key developments focused on faster imaging and delivery times are FFF x-ray beams, faster gantry speeds, and faster MLC leaf movements (Michiels et al. 2018). These could bring improved time efficiency for VMAT compared with IMRT (with gains from reduced beam on-time and MLC leakage especially for dynamic treatment delivery) and better plan quality for VMAT for both target coverage and OAR objectives (Michiels et al. 2018). Current VMAT plans are widely acceptable when delivered in a single arc, but studies have shown further clinical plan quality improvements if multiple arcs are used—a difficulty for platforms for which this would mean considerably increased delivery time (Michiels et al. 2018). A faster VMAT delivery time (e.g. through Halcyon's faster gantry speed) could enable multiple arc VMAT plans delivered in a similar time frame to conventional C-arm one- or two-arc solutions. A quicker clinical delivery could also reduce the significance of intrafractional motion (from internal anatomy changes) on geometric set-up and plan delivery within the fraction; an advantage already purposed by using VMAT over conventional IMRT and conformal therapies.

The new Halcyon system (see Figure 10.1) combines a 6 MV FFF standing-wave compact linac mounted in-line on an O-ring gantry with a maximum doserate of 800 MU per minute (when calibrated for 1Gy per 100 MU at dmax, 10×10 cm field and 100 cm SSD) (Cozzi et al. 2018; Li et al. 2018; Michiels et al. 2018; Varian 2018a) and a gantry rotation speed of about 4 rpm; four times faster than that permitted on C-arm linacs. A jawless MLC is utilised with a unique stacked and staggered leaf arrangement (Cozzi et al. 2018). The initial release of Halcyon enabled 1 cm MLC width field shaping and intensity modulation. The field shaping was achieved mainly by using the lower bank of MLC leaves (those closest to the patient) with the upper leaf banks tracking the leaf ends of the lower leaf bank and ensuring minimal interleaf leakage. Twenty-eight leaf pairs are configured in each opposing leaf bank (for both lower and upper banks) with a leaf-span of 28 cm and a maximum leaf speed of 5 cm/s (Cozzi et al. 2018; Michiels et al. 2018). The leaf banks have no carriages and there are no back-up jaws (in x or y

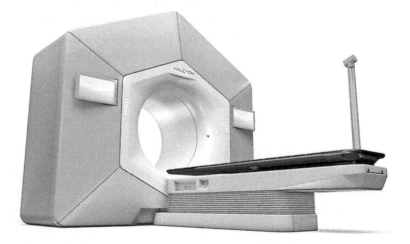

FIGURE 10.1 A photograph of the new Halcyon treatment delivery system.

Image courtesy of Varian Medical Systems.

directions); maximum field size is 28 × 28 cm at the isocentre (Cozzi et al. 2018), but there is no light field indicating field size, shape, or position on the patient (Li 2018). By comparison, the Varian Millennium MLC on the TrueBeam platform has a maximum gantry rotational speed of 1 rpm, a leaf width of 5 mm (in the central 20 cm portion of the field), a maximum leaf speed of 2.5 cm/sec and a leaf span of 15 cm. An integrated beam stopper means that radiation protection bunker (vault) design can be more straightforward and more compact (Varian 2018a).

10.4.2 First Version Imaging Options and Capability

The first clinically available version of Halcyon featured MV-only-based on-treatment imaging; using the main treatment beam and a flat panel imager-style EPID. This allowed the rapid use of a technology that had matured over many years, was tried and tested, and used the treatment isocentre. Imaging (and dose reconstruction from the treatment beam) both match the treatment geometry. All the facilities of 2-D MV on-treatment imaging together with 3-D MV CBCT are available to the user even though kV imaging, as already discussed in chapter 7, brings better subject contrast with generally a lower concomitant dose; MV imaging uses the treatment isocentre, can be used for in vivo dosimetry and adaptive therapy, acquires images of the delivered geometry (MLC field sizes, shapes and positions with respect to the patient's anatomy), and suffers less from artefacts from high Z materials (like hip prostheses) (Kirby et al. 1995; Herman 2005; Kirby and Glendinning 2006; Evans 2008; van Elmpt et al. 2008; Mijnheer et al. 2013; Held et al. 2016; Bedford et al. 2018; Cozzi et al. 2018; Zou et al. 2018).

Two principle imaging modes are permitted in the first version; 2-D static and 3-D volumetric imaging. MV-MV 2-D mode uses orthogonal imaging pairs (at fixed gantry angles of 0 and 90 degrees with 2 MU per image for a low-dose mode and 4 MU per image for a high-quality mode (Li et al. 2018). The MLC collimator angle is fixed at 0 degrees for on-treatment imaging and the maximum field size permitted is the maximum treatment field of 28 × 28 cm (Li et al. 2018). Volumetric imaging is achieved using MV CBCT through continuous gantry rotation from 260 degrees to 100 degrees; both full-fan and half-fan imaging modes are possible. Here, 5 MU are used for a low-dose mode and 10 MU for a high-quality mode with collimators again fixed at an angle of 0 degrees. The axial FOV is set at 28 cm, but longitudinally, the field can vary between 2 and 28 cm in 2 cm steps (Li et al. 2018). For general set-up, the patient is aligned first with a virtual isocentre indicated by the internal lasers within the system and then the couch moved into the centre of the wide bore (as in a CT simulator), MV image guidance is then used for set-up correction prior to treatment delivery. Imaging dose can be accounted for within Varian's own treatment planning system (Eclipse) during planning. For both imaging modes, the faster movement of the gantry is utilised so that Varian quotes a CBCT speed of acquisition as short as 15 seconds (Varian 2018a), verified independently by Michiels et al. (2018) at 14 seconds for a full-fan scan. Li et al. (2018) undertook a full dosimetric analysis of the on-treatment imaging options available for MV-based imaging.

Beyond the publications quoted above, initial investigations, testing and research has focused primarily on the prototype version with MV-only on-treatment imaging. For example:

- Early studies (Mihailidis et al. 2017a) that examined the MLC capability (in its 1 cm, per-leaf tracking mode using 6 MV FFF delivery) benchmarked against current, well-established TrueBeam Millennium 0.5 cm MLC for standard AAPM protocol (Ezzel et al. 2009) test plans for IMRT and VMAT using ion-chamber and standard pretreatment verification QA devices (e.g. MapCheck2 and ArcCHECK). Results showed the Halcyon MLC having a better pass rate than the TrueBeam MLC, thus comparing extremely well for both IMRT and VMAT under these test conditions.

- Kennedy et al. (2017) examined image quality parameters for the prototype AMFPI EPID using the MV treatment beam for its MV CBCT on-treatment imaging. Tests were performed using both imaging modes (low-dose and high-quality) using the Catphan 600 for different FOV field lengths, 200-degree fan arcs, and 12-second rotation time. Results compared well with both kV CBCT images from a TrueBeam system and TomoTherapy MV CT imaging and diagnostic CT when considering its use for on-treatment geometric verification. There was a closer comparison, naturally, to the MV CT imaging of TomoTherapy, but showing as-expected poorer contrast and spatial resolution compared with the kV imaging technologies. No improvements were observed with changes in FOV length, but better contrast linearity was seen for high Z objects, as one might anticipate for MV imaging (see chapter 7). In terms of practical delivery, Netherton et al. (2017) has shown that in simulated studies (for a range of treatment sites), IGRT times for image acquisition are reduced by 11 seconds and 49 seconds respectively for MV CBCT (3-D) and MV-MV (2-D) when compared to most conventional C-arm linacs.

- Scheuermann et al. (2017) examined the ability of the MV EPID for performing the standard Machine Performance Checks (MPC) (with its appropriate test tool for some checks) for critical daily QA checks, such as mechanical accuracy, radiation output/profiles, and MLC positioning/reproducibility for Halcyon with its in-line 6 MV FFF beam. Results were all within MPC tolerances and showed very good agreement with independent results with the exception of MLC reproducibility. All results were acceptable for daily clinical use.

- Since the initial version of Halcyon was only equipped with MV imaging modalities, concomitant dose is likely higher than that of kV imaging strategies, with also a lower image quality (primarily because of subject contrast, see chapter 7) compared with kV imaging. Early studies (Balter et al. 2017; Li et al. 2018) have examined the image quality (Kennedy et al. 2017) and the likely concomitant dose for both target and normal tissues doses (Balter et al. 2017; Li et al. 2018), and the ability of the associated planning system (Eclipse) to model and account for the imaging doses during the treatment-planning process. Results showed that for high-quality MV-MV imaging (2-D pairs), mean extra-target doses were about 1 cGy, and mean in-target doses about 2.5 to 4 cGy for major organs such as heart, lung, and spine (for isocentre set-up in the upper body) (Balter et al. 2017; Li et al. 2018). The extra-target measurements showed greater variation and dependency upon field size than in-target measurements. For volumetric imaging (MV CBCT), high-quality mode measurements revealed doses of about 6.5 to 8.5 cGy for heart, lung, and spine; theses were less sensitive to positioning of the treatment field, which is not too surprising, given the volumetric nature of the imaging and higher integral doses involved. Overall, the range was approximately 0.5 to 10 cGy for the range of imaging quality modes and techniques, with MV-MV techniques being approximately 50% of the volumetric MV CBCT techniques (Balter et al. 2017; Li et al. 2018). They also showed that the Eclipse system-modelled imaging doses were a maximum 0.5% different from those of ion-chamber measurements in phantom for overall prescription doses of 60 Gy (Li et al. 2018).

- Plan quality and efficiency (of both planning and delivery) has been studied in silico for a range of treatment sites (such as head and neck, lung, cervix, breast, prostate, and whole-brain treatments) and techniques (from standard 3-DCRT to IMRT, both dynamic and step-and-shoot, to multiple-arc VMAT) for the 1 cm MLC configuration of Halcyon (Anamalayil et al. 2017; Brady et al. 2017; Court et al. 2017; Mihailidis et al. 2017b; Netherton et al. 2017; Nitsch et al. 2017; Cozzi et al. 2018). In many cases, measurements have been confirmed through pretreatment QA and delivery in phantom (e.g. Anamaliyil et al. 2017; Mihailidis et al. 2017a, 2017b; Netherton et al. 2017; Nitsch et al. 2017). All studies showed that plan quality is comparable with those used clinically on conventional C-arm platforms (e.g. TrueBeam); with the potential improvements of using more arcs (e.g. for VMAT) within similar timescales of delivery and potential improvements in the plan quality (Riley et al. 2018). Delivery times from in phantom measurements are improved with the potential for delivering similar quality of treatments with a modest reduction in in-room treatment times (e.g. about 1 minute/patient) for patients (Netherton et al. 2017).

Now that the first patients have been treated on Halcyon (Penn 2017; Queens 2017; Leuven 2017; India 2018), further clinical results and studies will begin to emerge.

10.4.3 Continuing Developments

Further improvements are expected for this new technology, two of which are in terms of MLC width for planning and in on-treatment imaging capability (Cozzi et al. 2018; Varian 2018a). Initial studies have investigated Halcyon's plan capabilities using 1 cm wide MLC leaves, using per-leaf tracking (Cozzi et al. 2018; Varian 2018a). But since the system employs a stacked and staggered approach, there is the potential for 0.5 cm leaf width for planning and delivery by offsetting (in terms of leaf travel) the position of the upper leaves (Cozzi et al. 2018). This could bring further improvements in plan quality with the associated efficiencies of delivery exploited still further.

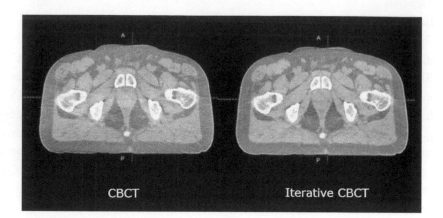

FIGURE 10.2   A sample image of the types of image-quality improvements possible with iterative CBCT reconstruction.

Image courtesy of Varian Medical Systems.

The second version of Halcyon now features kV on-treatment imaging (Cozzi et al. 2018; Varian 2018a) with all the associated benefits of imaging using kV energy x-rays (for both 2-D and 3-D imaging) in terms of subject contrast and contrast-to-noise ratio that one obtains with current kV systems on conventional C-arm platforms (see chapter 7). The arguments for MV imaging (in terms of reduced artefacts for high-Z implants, reduced equipment costs, and QA [Balter et al. 2017]), and employing the treatment isocentre itself for set-up correction) are still present, but kV imaging would bring likely image quality and concomitant dose benefits, with the improved efficiency and speed of imaging that Halcyon's technology and design affords. Coupled with improvements in kV image quality using iterative and statistical reconstruction (Wang et al. 2016; Chetty et al. 2017; Mao et al. 2017), image quality improvements are likely to be considerable compared to MV imaging, for considerably lower dose burden; akin to that already measured in other comparisons between MV and kV imaging (see, for example, Morrow et al. 2012 and chapter 7). An example of the scale of image-quality improvement expected is shown in Figure 10.2.

10.5 VERO

10.5.1 The Vero 4DRT System—Introduction

The Vero 4DRT is a highly innovative new hybrid radiotherapy platform that brings together a variety of treatment delivery and on-treatment imaging options into an O-ring technology format (Ouzidane et al. 2015b). First developed in Japan (through Mitsubishi Heavy Industries in conjunction with Kyoto University Hospital [Kamino et al. 2006, 2007; MHI 2012], and now marketed in combination with Brainlab AG (Feldkirchen, Germany), its design was unique in focusing on dynamic tracking (4-D) of target volumes as standard for radiotherapy. Built on an O-ring gantry, unique features included a rotating floor for the O-ring gantry (so that the gantry face need not be orthogonal to the couch for non-coplanar treatments), 2-D and 3-D kV and MV imaging facilities mounted on the gantry, an IR-marker detection system, and a gimbal-mounted C-band linac capable of a pan and tilt movements of the MV treatment beam (Kamino et al. 2006, 2007; Miyabe et al. 2011; MHI 2012; Ouzidane et al. 2015b; Miura et al. 2017). Treating its first patient in 2011 at Kyoto University Hospital, the company adopted the brand name from the Latin word *vero,* meaning truth.

The treatment delivery system consists of an in-line 6 MV C-band standing waveguide (about 35 cm in length and about 10 kg in weight) coupled with an in-built beam stopper, SAD of 100 cm and integrated 60 leaf (0.5 cm width) MLC producing a maximum field size of 15 × 15 cm at the isocentre with a maximum leaf speed of 5 cm/s (Nakamura et al. 2010, Depuydt et al. 2011, Solberg et al. 2014). The MLC is capable of full interdigitation and IMRT. The whole radiation head (about 600 kg in weight) is mounted on a gimbal with pan and tilt servos to provide a maximum excursion of the beam (at isocentre) of about 4 cm (equivalent to 2.5 degrees) with a maximum speed of about 6 cm/sec or 9 degrees/sec (Ouzidane et al. 2015b; Kamino et al. 2007; Depuydt et al. 2011; Solberg et al. 2014). Both static and dynamic modes can be employed for the gimbal mounting.

The O-ring gantry can rotate by approximately ±185 degrees at a maximum speed of 7 degrees/sec; the O-ring itself can rotate on its floor by ±1 60 degrees about the vertical axis at a maximum speed of 3 degrees/sec (Solberg et al. 2014). The O-ring gantry design exhibits excellent mechanical rigidity and movement, achieving an isocentre accuracy of 0.12 mm and 0.02 mm for gantry and ring (floor) rotations (Depuydt et al. 2011; Solberg et al. 2014). In addition, in static mode, the movement in the gimbal mechanism can compensate for any residual geometric distortion in the gantry, producing this excellent isocentric reproducibility for a full 360-degree gantry rotation (Kamino et al. 2006; Ouzidane et al. 2015b). The couch does not possess floor rotation capability; noncoplanar treatment beams are achieved through ring (floor) rotations (Kamino et al. 2007; Solberg et al. 2014; Ouzidance et al. 2015b). The couch top is a commercial robotic 6-DOF couch, as used on ExacTrac, which can help correct for lateral and longitudinal target volume rotations (Gevaert et al. 2012).

10.5.2 Imaging Options and Capability; Initial Results

The innovative Vero system has a range of image-guidance options using both ionising and nonionising radiation. The main x-ray options are mounted on the O-ring gantry and rotate with the treatment beam; the nonionising radiation system uses an IR camera mounted in the ceiling of the room, similar to that used on ExacTrac (chapter 7), enabling position monitoring in real time at no concomitant dose cost, and is independent of gantry position. Opposite the MV treatment beam is a traditional active-matrix flat-panel imager (AMFPI), and EPID for 2-D MV on-treatment imaging. Mounted either side of the lin ac-MV AMFPI axis are two kV x-ray tubes (orthogonal to one another) with associated AMFPIs in a cross-fire geometry through the patient with their imaging isocentre adjusted and calibrated to coincide with that of the treatment isocentre. Both MV and kV imaging options can be used in both static and real time (dynamic) modes, dependent upon concomitant dose burden. Through an approximately 225-degree rotation of the gantry, kV CBCT volumetric images (from either kV source) can also be acquired at 0.5-degree intervals (Solberg et al. 2014). Initial set-up is achieved using a pair of 2-D kV on-treatment images (De Los Santos 2013), a kV CBCT scan (Garibaldi et al. 2016), an initial patient localization using the IR camera system (Garibaldi et al. 2016), or a combination of these (Mukumoto et al. 2014; Akimoto et al. 2015). One of the key design objectives was intra-fractional imaging and, in particular, dynamic tracking of moving targets using the gimballed MV linac mounting with or without combined dynamic couch movements (De Los Santos et al. 2013). This is achieved using the pair of 2-D kV AMFPIs and the MV EPID panel, in consort with the IR imaging system. The reference images for 2-D imaging (kV and MV) are DRRs computed for the appropriate treatment beam and/or the imaging gantry angles for the kV x-ray tubes. KV CBCT volumetric images are compared with the pretreatment CT planning scans (De Los Santos et al. 2013). In terms of image quality, as we have seen in chapter 7, kV imaging brings better contrast resolution whereas MV offers better artefact reduction; 2-D kV imaging can be confounded by a lack of soft-tissue detail unless fiducials and other appropriate surrogates are used (De Los Santos et al. 2013).

For IR tracking, the position of the target is first identified through implanted fiducial markers and a 4-D predictive model is developed mapping the position of IR external markers with the internal fiducials in real time. During beam delivery, the future 3-D target position is calculated from the displacements of the IR markers using the 4-D model; the corresponding data is sent to the gimballed head for continuous tracking. The internal markers are also monitored, and where found to deviate outside predefined tolerances (e.g. 3 mm or more on three consecutive kV imaging frames), treatment can be interrupted and remodelling performed during the treatment session (Matsuo et al. 2014; Mukumoto et al. 2014). Experience with this 4-D technique has shown an excellent and stable tracking system, with some studies showing a predictive and tracking accuracy of less than 0.5 mm and average intrafractional errors less than 2 mm when using systematic prediction correction (Mukumoto et al. 2012, 2014; Akimoto et al. 2015). Long-term QA data shows excellent system stability for dynamic target tracking using these methods (Akimoto et al. 2015).

Depuydt et al. (2011) has demonstrated comparable performance of the 4-D dynamic tumor tracking (DTT) function with other clinical systems, using both phantom and patient studies; gimbal tracking for moving targets is very close to matching intended static dose distributions (Takayama et al. 2009; Mukumoto et al. 2012; Ouzidane et al. 2015b; Ono et al. 2018). Clinical use has been undertaken for sites such as lung/thorax, liver, brain, head and neck (HnN), ogliometastases, and spine using SBRT/SRS and hypofractionated procedures (De Los Santos et al. 2013; Solberg et al. 2014; Van Der Begin et al. 2016), mirroring how technology can change our practice, especially with SBRT (Aznar et al. 2018). Motion and tracking can be combined effectively for noncoplanar treatments under dynamic control, termed Dynamic Wave Arc (Burghelea et al. 2016, 2017). Initial clinical results (Matsuo et al. 2014) for lung patients are very promising, with target coverage little different from static techniques (Schwarz et al. 2017), but considerable reduction in normal lung volume dose (e.g. reduction in V20 of approximately 20%), and good toxicity results within a mean treatment time per fraction of about 35 minutes using DTT (Matsuo et al. 2014). Whilst more research is always needed (especially into imaging frequency, concomitant dose, tracking accuracy, and clinical outcomes), the initial accuracy and clinical results for DTTs using Vero seem very promising.

10.6 PROTON-BEAM THERAPY

10.6.1 Introduction

The clinical potential of proton-beam therapy (PBT) and charged particle (heavy ion) therapy has been considered since the proton was first discovered and the physical characteristics of dose deposition (particularly compared to x-rays) understood (Mayles et al. 2007; Jones et al. 2007; Degiovanni and Amaldi 2015, Durante and Debus 2018). As a clinical tool, interest has grown greatly over the last 10 years or so with a large increase in medical proton therapy installations worldwide (Mohan and Grosshans 2017). Its main advantages compared to a beam of x-ray photons is its relatively low entrance dose and its sharp rise at a specific depth (the Bragg peak) before a sharp fall to a near-zero dose, giving the characteristic clinical benefit of no exit dose. There is a definite range, depending upon the energy of the incoming protons, and a very large dose saving to distal normal tissues beyond the Bragg peak in complete contrast to photons (Mayles, et al. 2007; Jones et al. 2007; Blanchard et al. 2018; Degiovanni 2014). For this reason alone, the physical dose characteristics suggest it should be a superior form of radiotherapy compared to photons in terms of dose conformity around the target volume; more importantly are the potential for a much greater dose fall-off to surrounding normal tissues, lower doses to OARs, and lower integral doses (Mendenhall and Li 2012; Blanchard et al. 2018; Merchant 2013; Liao et al. 2008; De Ruysscher and Chang 2013).

10.6.2 Technology Changes

In this century, the technology has developed apace from large-scale changes to gantry-based systems (see Figure 10.3), a move from passive to scanning-beam systems, developments in cyclotron and synchrotron technology (especially in making them more efficient and more compact), beam energy selection, and on-board/in-room imaging systems (Mayles et al. 2007; Degiovanni and Amaldi 2015; Flanz and Bortfeld 2013; Degiovanni 2014, Schippers et al. 2018; Seco and Spadea 2015; Mohan and Grosshans 2017). The compactness of cyclotrons for PBT has meant that single-room and single-gantry systems are now being commercialised (Degiovanni 2014; Degiovanni and Amaldi 2015; IBA 2018; Mevion 2018). Developments continue researching into, for example, new types of particle acceleration (e.g. through dielectric wall accelerators and laser plasmas) and through more efficient examples of linear accelerator technology (Caporaso et al. 2008; Schippers et al. 2018; Degiovanni et al. 2016), again with consequent commercialisation for some (AdvancedOncotherapy 2018). The technological evolutions have improved on physical dose deposition and will change efficiency and cost effectiveness (Schippers et al. 2018); as the numbers of PBT facilities grow, comparisons between technologies are now being undertaken (Langner al. 2017). As we shall see clearly in the following sections, the full exploitation of these excellent physical characteristics of proton-beam therapy will not reach its full potential until there is greater understanding of proton-beam characteristics in vivo, and delivery is coupled with a secure, robust, and accurate form of on-treatment imaging, range verification, and an ability to adapt treatments (Flanz and Bortfield, 2013; Schippers et al.

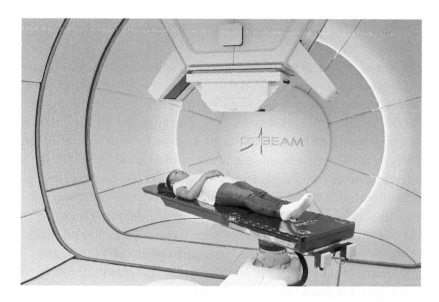

FIGURE 10.3 The Varian ProBeam PBT system with pencil-beam scanning and 2-D/3-D kV x-ray-based on-treatment verification imaging.

Image courtesy of Varian Medical Systems.

2018; Unkelbach and Pagenetti 2018; Hojo et al 2017; Seco and Spadea 2015). For the current technologies, Degiovanni and Amaldi (2015), Mohan and Grosshans (2017) and Schippers et al. (2018) have published excellent reviews.

10.6.3 Clinical Utility

As technology has stabilised and become more clinically useful, and as the number of facilities worldwide has increased, the clinical database has also developed in terms of experience and clinical outcomes for an ever-growing range of clinical sites. For example, these have included skull-based tumours (Ahmed et al. 2018), lesions of the eye (Mendenhall and Li 2012) and prostate (Ojerholm and Bekelman 2018), tumours of the head and neck (Frank et al. 2018; Kraan et al. 2013; Mendenhall and Li 2012, Blanchard et al. 2018; Romesser et al. 2016) and central nervous system (Mendenhall and Li 2012), sarcomas (Mendenhall and Li 2012), and tumours of the thorax (Mohan and Grosshans 2017; Liao et al. 2008; Engelsman et al. 2013; Ma and Lomax 2012; De Ruysscher and Chang 2013). A large body of evidence is growing particular for pediatric conditions   clinical cases where the physical dosimetric characteristics are perhaps most advantageous, considering patients whose tissues are still developing (and therefore more vulnerable to radiation-induced complications) and for whom integral dose is a vital factor for secondary cancer induction later in life (Merchant 2013; Mendenhall and Li 2012, Allen et al. 2012; De Ruysscher et al. 2012; Leroy et al. 2015). The May 2016 edition of the *Red Journal* (RedJournal May 2016) and the April 2018 edition of *Seminars in Radiation Oncology* (SRO Apr 2018) give perhaps some of the most complete recent clinical expositions for PBT. Modality comparisons (of protons versus photons) continue to show clear

physical dose advantages for PBT over photon solutions using IMRT and VMAT (Ahmed et al. 2018; Langendijk et al. 2018; Blanchard et al. 2018); but the clinical understanding is far complete, especially in terms of the relative biological effectiveness (RBE) compared to photons, and the effects of range uncertainties for the sharp Bragg peak (Paganetti 2014; Ilicic et al. 2018; Hojo et al. 2017; Mohan et al. 2017; Mohan and Grosshans 2017; Kirk and Kirby 2018; Degiovanni and Amaldi 2015). This is perhaps why more clinical trials are needed, in particular to add to a current paucity of data from which the clinical distinctiveness of PBT can be more conclusive (Allen et al. 2012; De Ruysscher et al. 2012; Merchant 2013; Leroy et al. 2015; Mohan et al. 2017; Mohan and Grosshans 2017). Two of the most comprehensive reviews (Allen et al. 2012; De Ruysscher et al. 2012) showed insufficient clinical evidence for effectiveness in sites such as head and neck, lung, GI, and even pediatrics (non-CNS)—no evidence for efficacy in hepatocellular and prostate tumours, some superiority for pediatric CNS, and some clear evidence of benefit for many ocular tumours. Prohibitive costs were seen as one of the major barriers in more facilities and therefore more clinical trials, which led to an overall conclusion that many more robust clinical trials and data were required to show any clear clinical benefit.

10.6.4 Clinical Challenges

One of the biggest clinical challenges to accurate and appropriate treatment delivery, so that the high-dose volume is delivered as planned within the patient, comes from the nature of proton-beam dose deposition (i.e. the Bragg peak) and internal anatomy changes either between planning and first fraction, or during the treatment course itself which, if unnoticed and unaccounted for, could lead to significant adverse effects (i.e. underdose of the target volume and over-dose of the OARs and nearby normal tissues) (Ma and Lomax 2012; Mohan and Grosshans 2017) (see Figure 10.4). Because of the nature of the delivery of dose along the proton beam's path, there is much greater sensitivity to anatomical changes within this volume (Zou et al. 2018). These anatomical changes could be due to weight loss/gain, organ filling or collapse, normal internal motion during treatment delivery (such as respiration and bowel movements) (Zou et al. 2018; Mohan and Grosshans 2017; Landry et al. 2015; Kraan et al. 2013; Hui et al. 2016). An important consideration is that it is not just the movement of the target volume that is of concern, but also the change in anatomy in the proton beam's path, which can have a considerable effect on dose distributions, especially where IMPT is concerned with possible interplay effects leading to hot and cold spots (Mohan and Grosshans 2017; Engelsman et al. 2013; Mohan et al. 2017). For many such motions, photon plans (even those with IMRT/VMAT) are more tolerant than protons in terms of effects on the dose distribution and target volume/OAR doses (Mohan and Grosshans 2017).

One of the clinical sites where such motion is most accentuated is for tumours within the thorax (Liao et al. 2008; Engelsmann et al. 2013; De Ruysscher and Chang 2014, Mohan and Grosshans 2017). Here, there can be a larger vulnerability to heterogeneities in the beam path from tumour and organ motion, anatomical changes in response to treatment, and lung collapse (atelectasis), thus presenting significant challenges to planning and delivering accurate and precise dose distributions. Tumour shrinkage can easily compromise

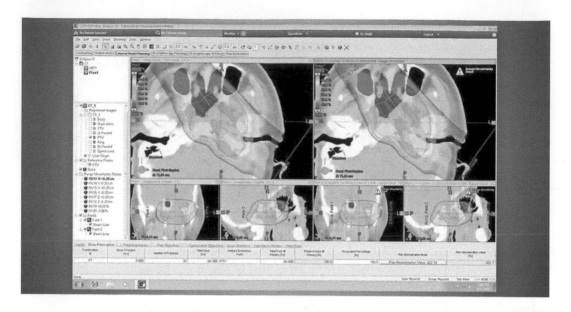

FIGURE 10.4 Varian Eclipse PBT treatment planning system, showing a typical PBT plan and the challenges associated with accurate dose delivery in the presence of potential internal anatomical changes for this head and neck patient.

Image courtesy of Varian Medical Systems.

the actual delivered dose distribution compared to the original plan; examples are seen whereby doses to organs at risk are considerably affected distally to the target volume if there is an associated decrease in the density of structures proximal to the target volume, which weren't the case at the time of treatment planning (Engelsman et al. 2013; Ma and Lomax 2012; Liao et al. 2008). To fully realise the potential of PBT, improvements are needed throughout the radiotherapy process.

One of these areas is within treatment planning and employing more robust strategies (Jones et al. 2007; Liu et al. 2015; Unkelbach and Pagenetti 2018). These include examining possible uncertainty scenarios of anticipated motion and the effects on the treatment plan. These could include shifts along orthogonal axes, range uncertainty, different breathing phases, tissue density changes etc. (Mohan and Grosshans 2017; Kirk and Kirby 2018; Unkelbach and Pagenetti 2018). In all cases, one would look to choose beam angles so that said anticipated changes minimised the risk of overdose to the nearby normal structures and reduction of target-volume dose (Engelsman et al. 2013, Mohan and Grosshans 2017; Romesser et al. 2016; Unkelbach and Pagenetti 2018, Kirk and Kirby 2018). Magnitudes of shifts should mirror target volume margins and range uncertainties employed for the plan. Review of the effects should reveal potential deficiencies and enable steps to be taken to ameliorate them (Mohan and Grosshans 2017). One such step is a robust on-treatment imaging strategy with suitable technology for detecting and correcting not just geometric shifts, but also anatomical changes that could compromise the dose plan (Jones et al. 2007, Schippers et al. 2018, Seco and Spadea 2015, Mohan and Grosshans 2017).

10.6.5 On-treatment Imaging

Because of the nature of dose deposition from the proton beam and the challenges identified above, the purpose of on-treatment imaging for PBT is primarily three-fold (Seco and Spadea 2015, Schippers et al. 2018). First, for geometric patient set-up as we have discussed throughout this book. Secondly, in helping to verify accurately the range of the proton beam because of the changes in vivo that can affect it, as well as the uncertainties in beam characterisation and treatment planning that affect our knowledge of it (such as dose algorithms, accurate stopping power determination, RBE and LET etc.) (Jones et al. 2007; Mayles et al. 2007; Zou et al. 2018; Unkelbach and Pagenetti 2018; Ilicic et al. 2018; Kirk and Kirby 2018; Mohan and Grosshans 2017; Mohan et al. 2017; Hojo et al. 2017; Degiovanni and Amaldi 2015). Thirdly, in terms of plan adaptation during treatment because of the potential consequences to the dose distribution and therefore the efficacy of the treatment, if internal anatomical changes occur and go unnoticed.

Compared to the on-treatment imaging capabilities for photon therapy, which is the main subject of this book, on-treatment geometric verification for PBT has been relatively primitive until recent years. Accurate and reproducible set-up has mainly relied upon excellent immobilisation and external set-up markers (skin tattoos, laser alignment, markers on immobilisation equipment etc.) with the use of 2-D radiographs and bony landmarks (Mendenhall and Li 2012). These 2-D radiographs, as we have for photons, are compared against DRRs from the planning CT scans for correcting set-up errors at the beginning of treatment fractions (Seco and Spadea 2015). This type of imaging has been central to PBT throughout its development for each delivered field and fraction (Schippers et al. 2018). However, to assure the quality of PBT plans during treatment delivery and exploit its dose delivery potential to the fullest, more sophisticated imaging is needed (Flanz and Bortfeld 2013).

Most newer PBT systems now come with a variety of on-treatment imaging options for geometric verification—a range of 2-D, 3-D and 4-D imaging techniques (Seco and Spadea 2015; Mohan ct al. 2017; Schippers et al. 2018; Pidikiti et al. 2018). Two-dimensional kV planar imaging is still available, either gantry-mounted and using flat-panel AMFPI technology either as a single x-ray tube and panel (which can take an orthogonal image pair, as on standard C-arm linacs, with a rotation of the gantry) or twin-mounted tubes and AMFPIs in a cross-fire technique through the isocentre, providing orthogonal imaging simultaneously (Seco and Spadea 2015, Pidikiti et al. 2018). Twin kV tubes and AMFPIs also may be mounted in the room (in a similar fashion to ExacTrac, see chapter 7) at floor and ceiling levels at approximately 60 degrees to each other and 45 degrees relative to the floor again in a cross-fire technique through the isocentre (Seco and Spadea 2015; Pidikiti et al. 2018). Reference images for set-up correction are 2-D DRRs from the planning CT scans reconstructed for the appropriate geometric view through the patient. Both bony landmarks and implanted fiducial markers can be used with planar images and/or fluoroscopic modes for real-time movement monitoring and tracking during treatment (Seco and Spadea 2015), again in a similar way to Exactrac.

Volumetric imaging is also now developing for PBT treatment rooms. For the gantry mounted kV imaging systems, this is through CBCT by gantry rotation and acquiring CBCT scans from one or both of the gantry mounted panels. This brings considerably better anatomical imaging to bear on-treatment, enabling anatomical changes to be spotted volumetrically on-treatment and plans adapted accordingly (Mohan and Grosshans, Engelsman et al. 2013; Ma and Lomax). Software used is very similar to that for photons, with manual or automatic bony-landmark or soft-tissue registration with respect to the 3-D planning CT scan set together with couch adjustments that can have six degrees of freedom and (Pidikiti et al. 2018). From this, daily on-treatment alignment can be achieved (if concomitant dose burden can be justified for that type of imaging protocol) with millimeter accuracy; couch and gantry systems are now capable of submillimeter, robotic precision (Seco and Spadea 2015; Mendenhall and Li 2012; Schippers et al. 2018; Degiovanni 2015, Ma and Lomax). Set-up time, like that for photons, can be a few extra minutes (Seco and Spadea 2015), but with considerable gain from the on-treatment image guidance and potential for necessary adaptation (see Figure 10.5).

Volumetric imaging can also be achieved through in-room CT systems (CT-on-rails) (Seco and Spadea 2015, Pedroni et al. 2011, Schippers et al. 2018) with the added benefit of identical 3-D/4-D high image quality to that of the planning CT scan set (with reduced scatter contribution compared to kV CBCT), but with the potential for patient movement between on-treatment imaging and treatment delivery (see chapters 7 and 8). Couch iso-centre movement, or "wobble" is less of an issue now with modern patient-support systems, especially those that are robotic as used for PBT and Cyberknife (Schippers et al. 2018). A very new development is an independent couch system with its own combined volumetric kV CBCT system (Schippers et al. 2018, MedPhoton 2018), ceiling- or floor-mounted on a

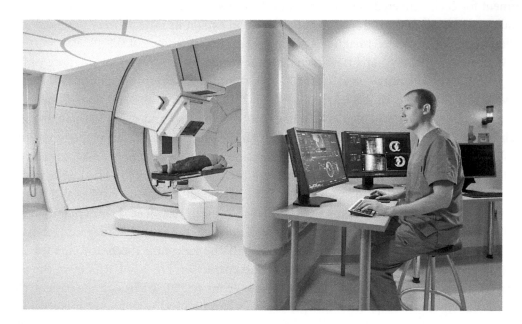

FIGURE 10.5 The Varian ProBeam PBT system with on-treatment imaging panels deployed (left) and during on-treatment verification imaging (right) with operator in position behind an (in-room) radiation protective screen.

Image courtesy of Varian Medical Systems.

highly accurate robotic arm. The kV x-ray tube and AMFPI are mounted on an "imaging ring" and can be used with different PBT facility arrangements, or even with photon therapy solutions (MedPhoton 2018).

Nonionising radiation technologies can also be used for on-treatment imaging, monitoring the external surface of the patient (with or without x-ray localisation) either directly or through surrogate surface markers (Seco and Spadea 2015; Fattori et al. 2017; Hoisak and Pawlicki 2018). Experience has been gained for thorax and chest-wall treatments; with results showing very good spatial resolution, a more efficient set-up procedure and good, intrafractional real-time monitoring capability without concomitant radiation dose (Fattori et al. 2017: Hoisak and Pawlicki 2018). For chest-wall treatments, the surface guidance has been found to be better than certain radiographic procedures (Hoisak and Pawlicki 2018).

With the development of MR for on-treatment verification (see section 10.7), it might be natural to consider its possible use here for proton and charged particle therapy as it brings with it very high-quality anatomical imaging volumetrically in real time with no concomitant dose burden (Liney et al. 2018, van Herk et al. 2018; Oborn et al. 2017). Volumetric, real-time MR imaging is already being shown as feasible and clinical on MR-guided linacs and Cobalt units (Green et al. 2018); the benefits would help realise the superior dose distribution advantages protons and other charged particles afford. Many of the cross modality interactions (i.e. the linac affecting the MR image quality, and the MR system affecting the linac and dose distribution) have been surmounted for photon therapy; although not completely and there are more on the list (Oborn et al. 2017). There would be similar challenges for PBT, with the added one being the therapy beam itself in air would also be affected by the MR's strong magnetic field since it is a beam of charged particles rather than chargeless and massless photons. That aside, many of the challenges were considered by Oborn et al. (2017) in a recent review paper, in which he and his colleagues considered a number of the potential challenges, and suggested technological configurations for both fixed-beam and gantry-based PBT with the particle beam being either perpendicular or in-line to the main magnetic field, as has been considered for different photon solutions (Kashani and Olsen 2018). With the added bonus of possible functional imaging, MR guided on-treatment imaging for particle beam therapy could be the next major advance, although there are many software and hardware processes to yet overcome.

With all of these methods, as is the case for photons, good quality assurance and routine calibrations and checks are vitally important, most notably for ensuring that the imaging systems can relate their frame of reference (their own isocentre, for example, in gantry-mounted kV CBCT) to that of the treatment isocentre. Techniques not using the treatment isocentre and a beam's eye view require rigorous assurance of the geometrical relationship between the two. For new facilities especially, end-to-end testing helps to eliminate systematic errors throughout the imaging process, and link the on-treatment imaging system to actual treatment delivery and the high-dose volume (Pidikiti et al. 2018). For partly this reason, system are still being researched into taking verification imaging using the proton or ion beam itself, so true beam's-eye-view geometric verification might be achieved, and real-time tracking of the tumour might be viable (Seco and Spadea 2015;

Martisikova et al. 2018; Johnson 2018). Johnson has published a recent review of the current state of the art for both proton radiography (pRAD) and proton-computed tomography (pCT), highlighting how the concept is decades old, still has considerable challenges to overcome before, say, pCT becomes a routine clinical viability (Johnson 2018; Depauw et al. 2014). In essence, the first requirement of pRAD is an energy of proton beam capable of passing through the patient. Particle detectors before and after the patient examine the change in range of the charged particles, caused by the water equivalent path lengths through the patient (Depauw et al. 2014; Johnson 2018). In this way, a range map through the patient can be reconstructed. Imaging dose would likely decrease, which would be especially good for pediatric cases (Depauw et al. 2014), and the high spatial resolution would suggest it could be a good tool for direct tumour tracking for, say, edge detection for lung tumours (Depauw et al. 2014; Johnson 2018).

In terms of on-treatment range verification, a number of approaches have been examined in this century (Buitenhuis et al. 2017; Knopf and Lomax 2013; Parodi 2015; Dendooven et al. 2015). The most investigated and widely available is based on PET imaging of the short-lived isotopes like 11C, 13N, and 15O, which are naturally produced in vivo from charged-particle nuclear interactions (Seco and Spadea 2015; Knopf and Lomax 2013; Jones et al. 2007). For protons, this originates from activation target elements such as oxygen and carbon within the body. The concept is relatively straightforward; where there is activation of these components producing positron emitters (whose subsequent 511 keV annihilation photons can therefore be imaged by a PET scanner), then that is where, geometrically, the proton beam has been deposited in reality in vivo. There is no external radioisotope tracer that needs administering to the patient—these isotopes are being generated inside the body through the proton beam's interactions (Zou et al. 2018). The timing of the imaging, however, is critical; since one must consider both the short and different half-lives of the activated elements and also biological washout through perfusion in the various tissues surrounding the initial point of activation (Seco and Spadea 2015; Dendooven et al. 2015; Parodi 2015: Zou et al. 2018). To avoid some of these problems, imaging methods must be combined with MC-based activity models to determine a more accurate range for the proton beam (Min et al. 2013; Fiorina et al. 2018; Knopf and Lomax 2013). Online imaging has been examined for measuring the predominant, short-lived (half-life about two minutes) 11C activity (Seco and Spadea 2015; Buitenhuis et al. 2017; Dendooven et al. 2015) and offline imaging focusing on the longer lived 15O activity (half-life about 20 minutes) (Seco and Spadea 2015; Knopf and Lomax 2013). Range verification accuracy has been achieved to about 1 to 2 mm for some head and neck studies (Seco and Spadea 2015). Various hardware configurations have been tested for the PET scanner within the treatment room (Zhu and El Fakhiri 2013; Min et al. 2013; AdvancedOncotherapy 2018), especially for on-line imaging, for these short-lived components and to minimise washout (Seco and Spadea 2015). Detector challenges are also present in terms of the speed of imaging required and avoiding detector "pile-up" (Helmbrecht et al. 2016; Buitenhuis et al. 2017; Dendooven et al. 2015). But even so, the millimeter accuracy is difficult for range verification using PET; a more accurate method utilises prompt gamma emission following PBT (Moteabbed et al. 2011; Lopes et al. 2015), which avoids some of the inherent PET imaging limitations. Here, depending upon the type of proton-nucleus interaction taking place, prompt MeV gamma rays can be emitted during treatment; their detection can produce a more accurate range estimation and has been used clinically (Xie et al. 2017; Richter et al. 2016). But for both these methods, the ideal is a combination with high-quality volumetric on-treatment imaging for registration range within the patient's anatomy. For internal anatomical changes (Liao et al. 2008; Hui et al. 2008; Rohan and Grosshans 2017), plan adaptation can then be performed with a knowledge of where within the anatomy the dose has actually been delivered in vivo.

10.7 MR ON-TREATMENT IMAGING

10.7.1 Introduction

As discussed in chapter 7, on-treatment imaging technologies which use nonionising radiation techniques have a distinct advantage in not contributing to the concomitant dose burden - and therefore can be used much more frequently (daily) for set-up correction and monitoring throughout the delivery of each treatment fraction. Daily and real-time on-treatment imaging is possible with such systems, and particularly useful for those clinical sites where intrafractional motion (target movement and deformation) is likely to occur during delivery (Bortfeld and Chen 2004; Yan 2010; Kupelian et al. 2008b; Jan et al. 2014; Hawkins et al. 2006). With its enhanced soft-tissue visualisation capability and full volumetric imaging capacity, MR systems would seem ideal for this purpose, with faster, cine sequences already showing acceptability for clinical studies of internal organ motion and deformation (e.g. Padhani et al. 1999; Ghilezan et al. 2005; Heijmink et al. 2009; Stam et al 2013). The potential gains are numerous, but it is an on-treatment technology and technique which is not without its challenges; for example with regard to health economics, adaptive strategies and algorithms, practical bore size, geometric distortion and linac calibration, workflow development, patient selection and staffing/training; to name but a few (Van Herk et al. 2018). From these studies alone, it is clear how on-treatment volumetric MR imaging could affect margin selection and/or the motion management and monitoring techniques used (Zou et al. 2018, Kashani and Olsen 2018); and could be an important tool for MR on-treatment guided adaptive radiotherapy (e.g. McPartlin 2016, Zou et al. 2018; Hunt et al. 2018; Choudhury et al. 2017; Pathmanathan et al. 2018).

10.7.2 On-treatment MR Guidance

Hybrid systems are being developed (and some used clinically) aiming to provide very high quality soft-tissue imaging (provided through MRI) for guidance during treatment delivery. The overall aim is a seamless integration of target imaging and treatment delivery with simultaneous, real-time operation for both (Menard and van der Heide 2014; Mutic and Dempsey 2014; Fallone 2014; Keall et al. 2014; Lagendijk et al. 2014; Lagendijk et al. 2016; Rankine et al. 2017; Wojcieszynski et al. 2017; Kashani and Olsen 2018). The challenges are to maintain the integrity of each (image quality and robust, accurate treatment delivery) whilst understanding the effect that they have upon each other, minimising the adverse effects as much as possible (Kirkby et al. 2008; Raaijmakers et al. 2008; Fallone et al. 2009; St Aubin et al. 2010; Santos et al. 2012; Liney et al. 2016).

Some of the key "interactions" for consideration are summarised in papers from researchers such as van Heijst et al. (2013), Liney et al. (2016), Mutic and Dempsey (2014), Fallone (2014), Lagendijk et al. (2014), Keall et al. (2014), and Lagendijk et al. (2016) and include:

- How the very strong magnetic fields from the MR affect the components and operation of the linac

- How the RF fields used for MRI can affect or interfere with linac components and the RF used for electron acceleration

- Practical limitations, in terms of linac mounting and associated components, and the MR imager construction

- How the RF used for electron acceleration in the linac waveguide might affect the RF pulses in the MR imager, and therefore the accurate reconstruction and formation of images

- How "breaks" in the MR magnetic field (because of the need to introduce linac components and the treatment beam itself) might affect MR field uniformity and therefore the MR image integrity (e.g. distortion) and quality

- How the strong magnetic field from the MR imager could affect dose deposition by exerting strong forces on secondary electrons produced after photon interactions in tissues and organs

- How MR images can be used for geometric verification, if they suffer distortion; or how they could be compared against x-ray-based reference images (e.g. CT planning scans)

The key designs integrate a small, compact linac into the midtransverse plane of the MR imaging coils or through/between the planes of open MR systems. Low (0.35 and 0.5 T) and high-field (1.0 to 1.5 T) MR imaging systems have been used thus far with a linac or three Cobalt sources mounted on a slip-ring gantry structure as the radiation delivery system (Mutic and Dempsey 2014; Fallone 2014; Keall et al. 2014; Lagendijk et al. 2014; Lagendijk et al. 2016; Rankine et al. 2017; Wojcieszynski et al. 2017; Kashani and Olsen 2018). Also proposed has been an MR system on-rails, which can be moved into the vicinity of the linac in the treatment room (Jaffray et al. 2014). The latest version of the MR-Cobalt system (which has had the most clinical usage so far [Fischer-Valuck et al. 2017]) now incorporates a compact, 6 MV in-line FFF linac in place of the three Cobalt sources (Mutic and Dempsey 2014). In terms of on-treatment imaging, the integrated MR system provides full volumetric images in real time, in both static and dynamic (real-time cine) modes for comparison with pretreatment planning CT slices, and for target volume/OAR tracking intrafractionally and daily inline adaptive treatments (Pathmanathan et al. 2018; Olberg et al. 2018; Green et al. 2018; Hunt et al. 2018).

Clinical sites such as CNS, head and neck, abdomen, pelvic tumours, etc. are anticipated as gaining most from such enhanced soft-tissue imaging (see Figure 10.6); together with better identification of associated organs at risk (Noel et al. 2015; Wojcieszynski et al. 2016; Acharya et al. 2016a; Liney et al. 2016; Bohoudi et al. 2017; Chen et al. 2017; Choi et al. 2017; Fischer-Valuck et al. 2017; Mehta et al. 2018; Ramey et al. 2018; Moghanaki et al. 2017). Clinical comparisons, from both in silico planning studies and from treated patients, have examined comparisons with current standards (e.g. VMAT on conventional linacs) for clinical sites anticipated to benefit from the improved soft-tissue visualisation that MR brings compared with kV x-ray based volumetric on-treatment systems. These have included breast (Acharya et al. 2016b), abdomen target volumes such as liver and pancreas (Glitzner et al. 2015; Wojcieszynski et al. 2016; Bohoudi et al. 2017; Ramey et al. 2018), oligometasta-sesoligomets in the abdomen and thorax (Henke et al. 2016), lung (Kim et al. 2017b), and SBRT type procedures (Choi et al. 2017; Kontaxis et al. 2017). Clinical plans are generally found to be comparable and acceptable compared to linac-based ones, but some studies call into question the MR-Cobalt system for finely targeted procedures, such as SABR for spine targets (Choi et al. 2017), and highlight MR-linac plans that can have greater similarity to conventional linac techniques in terms of conformity indices and critical structure constraints (Ramey et al. 2018). For all these initial studies, the possibilities of greater margin reduction for the PTV and therefore potential dose escalation/hypofractionation for

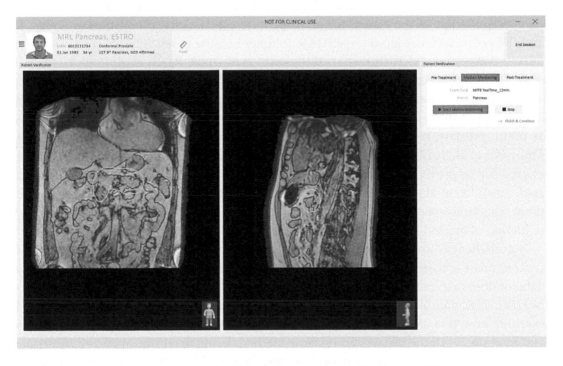

FIGURE 10.6   Typical images from the Elekta Unity MR-guided treatment delivery system, showing excellent soft-tissue image quality (abdomen in general and pancreas in particular) from this high-field system.

Image courtesy of Elekta.

certain sites are raised; and also 3-D/4-D on-line/in-line adaptive radiation therapy comes to a closer reality (Kupelian et al. 2008; Kupelian and Sonke 2014; McPartlin et al. 2016; Kashani and Olsen 2018; Yartsev and Bauman 2016).

10.7.3 Pretreatment MR and MR-Only Workflows

The importance and use of MR pretreatment for treatment planning and enhanced target volume/OAR delineation is already well known (Evans 2008, IAEA 2007, Khoo and Joon 2006), but it is based in a CT-based RT workflow through the use of MR-CT fusion etc. (Brock and Dawson 2014). The use for pretreatment simulation becomes heightened if one is considering an MR-guided on-treatment imaging platform as discussed above. In this way, reference images are of the same modality and quality as those acquired on-treatment, moving one toward a possible MR-only workflow (Torresin et al. 2015; Liney and Moerland 2014; Nyholm and Jonsson 2014). For this, the ideal MR pretreatment environment uses a wide-bore scanner, flat indexed MR compatible couchtops and immobilisation equipment, appropriate alignment lasers, close-proximity RF coils for improved SNR, and very good distortion correction algorithms (Torresin et al. 2015; Liney and Moerland 2014). At the pretreatment stage, good MR imaging brings with it T1 and T2 sequences that help differentiate between soft tissue for accurate tumour identification and staging (Zou et al. 2018), distortion correction calibrations and algorithms (Liney and Moerland 2014; Webster et al. 2009; Weygand et al. 2016) for ensuring acceptable geometric representation of the patient for accurate dose calculations, coregistration with CT, and the potential for reduced target/OAR volumes because of less interobserver variation during delineation (McPartlin et al. 2016; Datta et al. 2008; Rasch et al. 1999; Hentschel et al. 2011; Tanaka et al. 2011). The more advanced MR imaging modalities can be used, such as diffusion-weighted imaging (DWI) and multiparametric imaging, bringing both high-detailed anatomical and functional images into pretreatment planning for sites such as brain, prostate, and cervix (de Rooij et al. 2014; Guo et al. 2016; Torheim et al. 2017; Trebeschi et al. 2017). Real-time cine sequences can now help to provide temporal information for target-volume movement, internal organ motion, and deformation (Ghilezan et al. 2005; Liu et al. 2014; Uh et al. 2017; Hui et al. 2016; Fernandes et al. 2015), although there sometimes needs to be a trade-off between image quality (SNR) and the potential reduction of motion artefacts (Liney and Moerland 2014).

But challenges are still manifest for an MR only workflow for pretreatment and/or on-treatment scenarios. RT computerised planning relies upon accurate electron density data for dose calculations for MV energy x-ray treatments. A natural output from CT scanning, no such data is available from raw MR scans. Much research has been conducted to examine how this can be supplied for the MR-only workflow in pretreatment (Nyholm and Jonsson 2014); the two most promising methods being to outline, map and substitute electron densities for the purpose of dose calculations (Liu et al. 2014; Hsu et al. 2013; Johansson et al. 2011; Jonsson et al. 2010) or to attempt to convert MR image data (from specific sequences) directly into Hounsfield units (Korhonen et al. 2014; Kapanen and Tenhunen 2013). Additionally, whilst MR simulation may become more commonplace in the near future, the likelihood of MR on-treatment imaging on a large scale is as yet unlikely. Work is also being conducted to produce reference images from MR pretreatment simulation that can be compared with on-treatment x-ray-based images for 2-D use (as a substitute for DRRs) (e.g. Yu et al. 2014, Price et al. 2016) and for 3-D use (as a substitute for CT planning scans) (e.g. Price et al. 2016).

10.8 PET-MR

10.8.1 Introduction

In chapter 7, we examined the technicalities of PET-CT and its use in radiotherapy for the purpose of on-treatment verification imaging. One of the newest developments moving on from this is to naturally examine the possibility of integrated PET-MR imaging, with MR bringing all of the benefits of superior soft-tissue delineation at no dose burden to the patient, thus improving the metabolic-anatomic imaging capability pretreatment (Buchbender et al. 2012; Becker and Zaidi 2014; Jadvar and Colletti 2014). Much research is on-going into the anticipated clinical gains from PET-MR of more accurate staging pretreatment, but also improved follow-up by better differentiation of scar tissue from recurrence (Quick 2014; Becker and Zaidi 2014). PET-CT added functional imaging to high-quality anatomical imaging for the purpose of initial diagnosis, staging, and pretreatment computerised planning, and with it the inherent accurate registration of the two modalities by virtue of being a combined, single machine. An integrated PET-MR machine would bring the same registration accuracy, but some greater technological challenges have needed to be surmounted (and still need further work) in developing this technology. All of the developments of PET itself, in aiding diagnosis and staging (Jadvar and Colletti 2014; Valk et al. 2006; Thorwarth and Alber 2010; van den Brekel et al. 1990), investigating alternative tracers to FDG (Grosu and Weber 2010), improving target volume delineation for clinical sites such as head and neck, GI, cervix, lymph, bone, and soft tissue (Malapure et al. 2016; Sonni and Iaguru 2016; Tabacchi et al. 2016) with reduced interobserver variability would be expected here and be improved further because of the enhanced anatomical spatial resolution and definition of MR in the majority of cases, especially compared to PET alone (Valk et al. 2006).

The setting here in describing PET-MR has two purposes; one, to inform about this new technology in the pretreatment setting, and two, to understand the reference datasets it would offer for on-treatment verification imaging. Since pretreatment imaging informs of the desired treatment plan and geometric positioning of the high-dose volume, it is important to understand what PET-MR could offer alongside its limitations.

10.8.2 Initial Investigations and Challenges

Although the concept of combining PET and MR imaging systems into one integrated unit is easy to visualise, making it a reality is not so easy. The cost involved in developing MR-based prototypes is considerable compared to CT-based platforms. Like PET-CT, the onsite provision of radionuclides must be considered for the relatively short half-lives required for in vivo use. Perhaps one of the largest challenges involved was changing the PET detector technology. Previous standard technology made use of scintillator crystals attached to individual photomultiplier tubes for the detection of the annihilation gamma ray photons exiting the patient. However, the cascade of free electrons within the vacuum of the photomultiplier tube are highly susceptible to the influence of strong magnetic fields. For the detectors to work required the development of avalanche photo diodes (APDs)— solid-state devices able to detect gamma ray quanta even in magnetic fields up to 9 T and still efficiently convert the scintillation light photons into electrical signals (Quick 2014). The challenges of Hounsfield unit generation and MR geometric distortion, as discussed in sections 7.4, 8.2 and 10.7, are important ones for any radiotherapy planning and verification procedures (Quick 2014; Van Herk et al. 2018). MR-based attenuation correction for the PET signals is another challenge, together with the need to reduce the effects of the PET system on the MRI environment (e.g. to limit its effect on the uniformity of the static MR field, the gradients fields, or RF emission and reception) and reduce the effects of the PET system disturbing the electromagnetic MR fields and signals (Delso et al. 2011; Quick 2014).

Early investigations examined PET and MR in combination, using PET-CT as a gold standard for comparison (Drzezga et al. 2012; Buchbender et al. 2012); first clinical work-ups included separate MR and PET imagers in one room (Quick 2014; Jadvar and Colletti 2014) and prototype PET inserts designed to fit inside the bore of the MR unit to accommodate head and neck imaging (Hong et al. 2013; Jadvar and Colletti 2014). The natural development was then of a fully integrated, hybrid, whole-body PET-MR system. One of the first to be commercialised was that of Siemens (Biograph mMR, Siemens AG, Erlangen, Germany), enabling simultaneous PET/MR data acquisition (Quick 2014; Delso et al. 2011; Drzezga et al. 2012; Jadvar and Colletti 2014). This particular system consists of a 3 T whole-body MR unit with a length of just under 2 m (magnet length about 1.6 m), hosting a fully integrated PET detector assembly at the isocentre and providing a 60 cm patient bore (Quick 2014; Delso et al. 2011).

The advantages of such a whole body system include:

- Functional information not possible from CT, but which could complement functional and multiparametric MR imaging

- Tissue definition superior to that of CT

- Inherent accurate registration of the two imaging datasets because of the simultaneous data acquisition, without patient repositioning between scans

- Improved pre- and post-treatment assessment

- Potential for dose escalation (because of more accurate delineation and therefore smaller target volumes) and even spatial dose modification within the target volume, based upon, for example, higher tracer uptake

- Reduced concomitant dose burden compared with PET-CT alone

The difficulties include:

- Greater cost envelope compared with PET-CT technology

- Greater considerations for networking and data handling, especially with regard to import of image data into treatment planning systems for computerised treatment planning

- Acceptable distortion correction and Hounsfield unit/electron density calculation/ mapping (if possible)

- The need for further image fusion with planning CT for computerised dose calculations and, therefore, also the need for an added scanning procedure for the patient.

10.8.3 Clinical Utility

Clinical utility has been examined for sites such as head and neck, chest, abdomen, brain, and pelvis (Buchbender et al. 2012; Drzezga et al. 2012; Becker and Zaidi 2014; Szyszko and Cook 2018; Jadvar and Colletti 2014) as well as some pediatric conditions (Kirchner et al. 2017). Advantages have been demonstrated clinically for a number of situations (Becker and Zaidi 2014). For example, a diagnostic advantage of using MR when CT images are artefact-laden for scan volumes containing high Z components (dental fillings, hip prostheses) include improved soft-tissue delineation for head and neck (Szyszko and Cook 2018; Becker and Zaidi 2014); chest, pelvis, and abdomen imaging (Buchbender et al. 2012); improved follow-up and recurrence monitoring for metastatic disease (Quick 2014); and the potential for using different radioisotope tracers (e.g. Carbon 11) for nonmalignant conditions such as Alzheimer disease (Quick 2014). Clinical problems have also been encountered in terms of MR imaging motion artefacts and apparent changes in geometry for different MR sequences (Becker and Zaidi 2014), thus leading to observations of higher activity in different anatomical locations. Multiparametric imaging (e.g. DWI) becomes possible with MR, bringing added functional information to bear for treatment planning and the definition of target volumes. Some studies, however, have shown that the PET component (within PET-MR imaging) is still a valid addition over functional whole body MR alone (e.g. for staging paediatric lymphoma patients, even when the PET/MR is unenhanced or without added DWI information) (Kirchner et al. 2017). In addition, the better spatial resolution of CT over MR, still makes PET-CT a better choice for its higher sensitivity in clinical situations, such as imaging lung metastases, but with the required justification for the likely increased concomitant dose from PET-CT compared with PET-MR (Buchbender et al. 2012: Szyszko and Cook 2018).

10.8.4 Imaging Capability

In chapters 7 and 8, we discussed the principles of understanding for pretreatment imaging in bringing better information to bear for target volume delineation, and OAR definition. Here, PET-MR would bring better anatomical and functional information for these purposes in informing the treatment plan, with likely reduced user outlining variability and smaller target volumes and margins (Hidegéty et al. 2015; Leclerc et al. 2015). The target volume as a whole would be more precisely informed, and areas of higher tracer uptake could inform the planning of intensity modulation for better conformity around the target volume and also identify regions within the PTV for higher treatment doses because of probable radio-resistance/hypoxia (Aerts et al. 2010; Van Elmpt et al. 2012). None of these aspects are without their challenges (Geets et al. 2007; Lin et al. 2007; Callahan et al. 2013); segmentation alone (i.e. accurately identifying the volume of tracer uptake and its relationship to viable tumour cells) is still an on-going debate (Geets et al. 2007; Callahan et al. 2013).

However, on-treatment geometric verification is needed to secure the safe use of such smaller margins. For this purpose, we might envisage PET-MR imaging information being fused with a CT planning scan for computerised treatment planning, especially if patients could not have a PET-MR scan in an appropriate treatment position with associated immobilisation equipment. The CT scan data would then provide the appropriate reference images for on-treatment imaging (2-D DRRs and 3-D CT scans), provided the appropriate soft tissues and/or fiducials could be related clearly on PET-MR and CT scans.

As discussed in chapters 7, 8 and earlier in this chapter, the MR data from PET-MR could form a reference dataset of its own, if the likely distortion was adequately corrected; and dose computation could be achieved by appropriate methods (see section 10.7.2). It could then likely form the basis of an MR-only workflow, providing MR scans for reference on-treatment against, say, an MR-guided treatment technology (see section 10.7.3). Or one would need to find ways of converting MR image data into pseudo-CT data for creating appropriate DRRs and/or CT scans for 2-D, 3-D and 4-D on-treatment verification (Yu et al. 2014; Price et al. 2016).

10.9 EMISSION-GUIDED/BIOLOGY-GUIDED RADIOTHERAPY (EGRT/BGRT)

10.9.1 Introduction

As we have seen in chapters 7 and 8, the advantages of functional imaging, in being able to molecularly target tumour cells by their metabolic pathway through PET and other functional imaging modalities, are numerous. To be able to refine the target volume to concentrate high therapeutic dose in areas likely to contain concentrations of actual tumour cells, rather than through purely a visualisation of gross anatomy, is a powerful tool. Its use for pretreatment imaging, usually in process or technological combination with CT and MR, has been highlighted in aiding diagnosis, staging, and treatment planning. A natural extension of this is to introduce its use into the on-treatment scenario for either biological geometric verification of the target volume, an aid to plan adaptation during treatment or for helping to guide and pinpoint, biologically, the treatment beams themselves directly to concentrations of tumour cells. Using PET as the imaging modality for on-treatment guidance requires understanding of all the difficulties, opportunities and challenges associated with PET imaging (as we have outlined in sections 7.4 and 8.2), but provides a unique potential for a treatment delivery technology. It is a technique now being termed emission-guided radiotherapy (EGRT) or biologically guided radiotherapy (BGRT).

10.9.2 Initial Design and Clinical Utility

As with the previous technologies, integration of the imaging and therapy platforms are the ideal, so one single technological system is used for the patient for both static and dynamic on-treatment imaging and treatment delivery (Han et al. 2018; Seyedin et al. 2015; Yang et al. 2014; Fan et al. 2012, 2013). Traditional x-ray-based on-treatment imaging is used for obtaining good anatomical set-up geometrically, but then the radiolabeled tracer, once injected and accumulated into the target (tumour) cells, makes these cells a type of biological fiducial markers for inherently tracking motion and correcting set-up uncertainties (Phillips et al. 2018).

Initial prototyping and feasibility studies examined these possibilities; key objectives being to use the PET gamma ray emissions in real time for direct tumour-cell tracking during treatment delivery (Ishikawa et al. 2010; Fan and Zhu 2010, Yang et al. 2011). Treatment beams or beamlets would be directed to the target volume along the lines of response (LOR) determined from the gamma ray emissions and coincidence counting for PET (see sections 7.4 and 8.2). A fast rotating ring gantry could be used for the in-line linac and imaging, alongside a ring structure of PET detectors.

The preproduction system consisted of a compact 6 MV in-line linac with a binary MLC (similar to that on tomotherapy, section 10.1), ring mounted PET detectors, kV CT imaging and MV imaging flat panels. All are combined onto a slip-ring gantry structure that can rotate around the patient at approximately 60 rotations per minute as the patients is translated through the bore of the system (again, like Tomotherapy) (Han et al. 2018; Partouche et al. 2018; Fan et al. 2013; Yang et al. 2014). Dose delivery and modulation is achieved through the 64-leaf binary MLC, with each leaf having a nominal width of 0.625 cm at the isocentre (Liang et al. 2018; Fan et al. 2013). An axial delivery mode is used, so the couch is translated in small steps and intensity modulated beams fired back along the PET lines of response over many gantry rotations while the couch is stationary; treatment delivery is axial rather than helical (Liang et al. 2018; Fan et al. 2012, 2013; Yang et al. 2014; RefleXion 2018). Initial clinical set-up, on-treatment imaging verification, and delivery is envisaged along the following pathway (Fan et al. 2013; Han et al. 2018; Yang et al. 2011; RefleXion 2018):

- Patient is administered PET tracer and waits for a time to allow optimum uptake, as per a standard PET protocol

- Patient is positioned on the couch and set-up corrections made using an on-treatment localisation kV CT scan and/or MV CBCT imaging protocol; reference is a planning CT scan set

- A short duration PET/MVCT (if used for alignment) scan is acquired for alignment and calibration, and additionally to update the planning map

- Treatment is delivered following LOR detection (and qualification) and beamlet delivery back along the valid LORs. The intensity modulation is achieved following a planning map, updated and modified by LOR detection

Early work featured proof of concept and feasibility studies (Ishikawa et al. 2010; Fan and Zhu 2010; Yang et al. 2011), with early simulated studies for lung and prostate sites (Fan et al. 2012). Tracking potential was examined for treatment simulations and planning for breast metastases and using dynamic lung phantoms (Fan et al. 2013, Yang et al. 2014). Planning studies investigated plan quality compared with VMAT and PBT for the thorax (Seyedin et al. 2015), exhibiting similar plan qualities and a potential reduced geometric uncertainties on-treatment and a reduction of doses to proximal critical structures. More recent papers focused on the capability of the associated treatment-planning system (Liang et al. 2018; Partouche et al. 2018) for prostate, esophagus patients, and dosimetric evaluations and comparisons for nasopharyngeal malignancies and oligometastases from prostate cancer patients (Han et al. 2018; Partouche et al. 2018; Phillips et al. 2018). Single-isocentre techniques were shown to be clinically feasible for multiple sites and good, sharp dose gradients were exhibited around the PTV, particularly for SBRT/SABR type procedures (Phillips et al. 2018; Partouche et al. 2018).

Recently (October, 2018), the system was launched commercially at the American Society for Radiation Oncology (ASTRO) annual meeting through RefleXion Medical as a fully integrated biology-guided radiotherapy system (ARO 2018; RefleXion 2018).

10.10 ADAPTIVE RADIOTHERAPY

10.10.1 Introduction

The subject of adaptive radiotherapy (ART) was first introduced by Di Yan (2010) and has been the subject of intense research and development ever since. The concept again is relatively straightforward; we know that the CT planning scan is only a snapshot in time of the patient and the computerised treatment plan based upon it is likely to be different when delivered unless the patient's size, shape, position etc., are identical to the planning scan, which is highly unlikely. The magnitude of the difference between planned and delivered dose will vary depending upon the significance of the anatomical changes. But some changes are always likely because of the time between initial planning scan and treatment starting; and also because internal organ motion can have a significant effect, depending upon the anatomical site (e.g. rectal and bladder filling/emptying for pelvic patients). The differences can be interfractional and/or intrafractional, and can also be due to likely changes during the treatment course; i.e. weight gain/loss, tumour shrinkage in response to the treatment itself or tumour growth, or atelectasis for treatments involving the thorax (Kashani and Olsen 2018; Dieterich et al. 2016b; Zou et al. 2018; Liao et al. 2008).

The development and use of on-treatment verification imaging (2-D, 3-D, and 4-D) has enabled two main aspects here: (1) readily and accessibly revealing the magnitude of these anatomical changes at the start of and throughout the treatment course (changes that could only have been observed previously by repeat CT or MR scans at key points during the treatment course); (2) decision protocols for plan adaptation (replanning in response to changes) off-line or even on-line, given the availability of on-treatment imaging and planning software at the treatment unit. Although modern on-treatment imaging has enabled easy geometric correction of set-up error, only certain types of changes to the target volumes can be accounted for by couch translations and/or rotations (through appropriate couch control and technology). For example, changes in the target volume shape (perhaps due to nearby organ filling/emptying) are not necessarily accounted for by simple translations and rotations; nor too a change in target volume in response to treatment or the independent motion of volumes/organs within the target volume as a whole (Kainz et al. 2017; Yan 2010; Zou et al. 2018). For this, adaptation of the plan is necessary by deformation (warping) and mapping of the organs between their position on-treatment and those of the original planning scan in relation to the expected dose distribution. The development of deformable models and algorithms is still one that needs more research (De Los Santos 2013; Wang et al. 2008). Ultimately, the key focus is the dose that has been delivered to parts of the target volume that should inform the decision making (not just geometry) estimated by warping and matching tissues between on-treatment images and the pretreatment planning scan and dose distribution (from a predicted dose). The ideal adaptive strategy is one that is planned into the treatment process (not just reacting to volume changes) with daily adaptation and dose guidance (in 4-D); the dose is calculated from that which is measured volumetrically in vivo, rather than from a TPS prediction (Yan 2010; Sonke 2010; Olivera and Mackie 2012; Sharpe et al. 2012; Chen et al. 2006; NRIG 2012). At present there are no national guidelines for standard adaptive or for dose-guided radiotherapy; some adaptation is advocated as part of national trials, but it is still very much work in progress (Kashani and Olsen 2018; Dieterich et al. 2016b; NRIG 2012).

10.10.2 Adaptive Radiotherapy Performed Off-line

Most experience to date (either through simulation, planning, or clinical studies) is based upon off-line approaches (Kashani and Olsen 2018; Dieterich et al. 2016b; Zou et al. 2018). Here, daily on-treatment imaging (usually volumetric) is used to trigger a decision to adapt the treatment plan; the replanning being undertaken off-line following the same workflow process as that for the initial planning. The patient might be resimulated (CT simulated), and the planning process followed as before with a timescale of a few hours to perhaps a few days. For most patients, this is a single episode that may be sufficient clinically, but is also partly due to the resource implications (Kashani and Olsen 2018). These offline responses are a useful and valuable form of adaptive radiotherapy, made possible from the availability of on-treatment volumetric imaging in particular; but some anatomical changes occur in a much shorter timeframe requiring an appropriate plan whilst the patient is on the treatment couch, for example, interfractional motion and deformation of the target volume for pelvis and abdominal clinical sites (e.g. Hafeez et al. 2017; Kong et al. 2018a; Bradford and Kirby 2017; Uh et al. 2014; Olberg et al. 2018; Heemsbergen et al. 2007) where soft tissue changes are likely day-to-day affecting the shape of the target volume and/or the proximity of OARs to the high dose volume.

10.10.3 Adaptive Radiotherapy Performed On-line

For the shorter temporal changes mentioned above, an off-line strategy with an adaptive plan for delivery, the subsequent fraction is likely to be ineffective. For this, an on-line strategy is better—one that is defined as the process of modifying the plan (or applying a modified plan), based on the on-treatment image data, and delivered during a given fraction whilst the patient remains on the treatment position (Kashani and Olsen 2018; Yan 2010). One such strategy, which has been used with good effect for soft-tissue targets like the bladder, is known as plan-of-the-day. In it a plan is chosen, based upon information in the volumetric on-treatment image, as the best and most appropriate plan for that day, given the anatomical changes that occurred (Murthy et al. 2016; Hafeez et al. 2017; Kong et al. 2018a, 2018b). This has been used in a number of clinical sites (discussed later).

An ideal strategy is proactive, using data from that particular fraction and also from previous fractions to correct for any imperfections in accumulated dose delivered so far; the plan is then modified for that fraction—a type of dose-guidance technique. Work is progressing to try to make dose modification on-line a reality (i.e. to replan/modify dose delivery from the on-treatment image) with as smooth, efficient, and as safe a process as possible achieved in minutes rather than hours (Kashani and Olsen 2018). This work might only be realised to its fullest potential with the use of MR on-treatment image guidance (Olberg et al. 2018; Green et al. 2018; Ramey et al. 2018; Hunt et al. 2018; Choudhury et al. 2017; Pathmanathan et al. 2018) for observing and potentially tracking target volumes in real-time intrafractionally with no concomitant dose burden. For this gold-standard, the online adaptive process must include dose calculations on on-treatment image sets, deformable modelling, automated contouring, fast dose calculation, and optimisation and verification of the modified plan—all in a few minutes (Kashani and Olsen 2018). Add in daily in vivo dose measurement and on-treatment 4-D imaging (where necessary) together with proton-beam therapy and biological dose modelling/optimisation, and the goals of very high-dose conformity to the target and minimisation of OAR doses in terms of clinical effects would take another major step forward for radiotherapy (Chen et al. 2006; Sovik et al. 2010; Mohan et al. 2017; Mohan and Grosshans 2017; Zou et al. 2018). With proton-beam therapy (as we have seen earlier in this chapter), the need for adaptation may be greater still for certain clinical sites because of the effect on the dose distribution by internal anatomical changes between the original plan and the situation following on-treatment imaging when treating with charged particles (Mohan and Grosshans 2017; Mohan et al. 2017; Liao et al. 2008; Kraan et al. 2013; Hui et al. 2008; Engelsman et al. 2013; Ma and Lomax 2012).

10.10.4 Experience With Different On-treatment Imaging Technologies

Adaptive radiotherapy is being developed, either through planning and feasibility studies or through real clinical experience, on a number of on-treatment volumetric imaging platforms. For example using:

- kV CBCT on Varian and Elekta linacs (Hafeez et al. 2017; Chen et al. 2014; Foroudi et al. 2014; Bradford and Kirby 2017; Duffton et al. 2016; van den Bosch et al. 2017; Berkovic et al. 2017; Kong et al. 2018a, 2018b)

- MV CBCT on Siemens linacs (Sonke 2010; Castadot et al. 2010; Chen et al. 2006)

- In-room CT from Siemens systems (Wang et al. 2008; Ahunbay et al. 2016; Wu et al. 2008)

- MVCT from TomoTherapy (Chen et al. 2014; Kainz et al. 2017; Murthy et al. 2016; Woodford et al. 2007)

- MR on-treatment image guidance from Viewray, Elekta etc. (Kupelian and Sonke 2014; Kontaxis et al. 2015, 2017; Vestergaard et al. 2016; Pathmanathan et al. 2018; Choudhury et al. 2017; Hunt et al. 2018; Oborn et al. 2017; Kashani and Olsen 2018; Olberg et al. 2018; Green et al. 2018; Ramey et al. 2018; Mutic and Dempsey 2014; Hu et al. 2015)

Most current experience is gleaned from the work using kV CBCT on standard linacs and the use of TomoTherapy (MVCT). For TomoTherapy (as seen earlier in this chapter), on-treatment is a necessary and integral part of daily use for treatment. It also utilises the treatment beam itself (albeit with a slightly modified MV energy of x-rays), and therefore the potential for in vivo dose measurement and dose guidance is readily accessible as a system. For kV CBCT, the concomitant imaging doses and image quality are better (lower dose, higher quality) than those of the MV systems, but requires extra quality assurance to ensure the imaging and treatment isocentres coincide. They would not be able to measure delivered dose in vivo, but one could combine the anatomical information from the kV CBCT images with dose information measured from the MV EPID for use in a dose guidance strategy (Bedford et al. 2018; Spreeuw et al. 2016; Rozendaal et al. 2014).

10.10.5 Adaptive Radiotherapy in Clinical Use

Clinical examples are numerous. They include:

Head and neck sites (Castadot et al. 2010; Kainz et al. 2017; Chen et al. 2014): Some studies (using both kV CBCT and MVCT) have compared treatment with and without replanning and found that not all patients would benefit from such replanning, but significant benefit has been found for selected patients (Chen et al. 2014).

Lung and thorax sites (Sonke 2010; Duffton et al. 2016; van den Bosch et al. 2017; Dieterich et al. 2016b; Berkovic et al. 2017): This is perhaps one of the most challenging sites, with likely anatomical changes and target volume movement, too. Some tumours have been found to respond during treatment, perhaps shrinking to approximately 50% of initial volume, sometimes asymmetrically, in which case geometric set-up on the target volume only could mean surrounding tissues are in quite a different geometry compared to the original plan (Dieterich et al. 2016b; Zou et al. 2018; Woodford et al. 2007). When adaptive therapy is applied, some significant OAR dose reduction has been noted with an improved outcome for those with a higher tumour shrinkage gradient (Berkovic et al. 2017); but also showing that adaptive strategies are not suitable for all patients (Duffton et al. 2016). Atelectasis (lung collapse) is a condition with which one must also contend for thorax treatments; studies showing that this is easily identifiable with kV CBCT (Duffton et al. 2016) with the possibility of automatic analysis, and selection of patients for replanning and adaptive therapy, based upon volumetric image-density change analysis between the planning and CBCT datasets (van den Bosch et al. 2017).

Rectum and cervix (van Wickle et al. 2017; Thornqvist et al. 2016; Bradford and Kirby 2017; Uh et al. 2014; Stewart et al. 2010): Rectum patients are challenging for their geometric verification alone, especially when intensity modulated techniques are used (Lowe et al. 2016). Because of difficulties with soft-tissue movement and matching, adaptive strategies have tended to use an offline approach with better sparing of nearby OARs (e.g. bladder) compared to non-ART approaches. For cervix patients, a CTV-PTV margin of 15 to 20 mm is often required because of the internal organ motion both inter- and intrafractionally from bladder and rectum changes (Lim et al. 2011). Soft-tissue matching is challenging especially for IMRT/VMAT treatments; adaptive approaches have used margin or plan-of-the-day strategies (Ahmad et al. 2013; Heijkoop et al. 2014; Bradford and Kirby 2017). Cervix may be one of many sites that would benefit from MR on-treatment imaging and plan adaptation (Uh et al. 2014).

Renal and pancreas (Kontaxis et al. 2017; Olberg et al. 2018; Green et al. 2018; Ramey et al. 2018): These are two clinical sites also likely to benefit most from MR on-treatment imaging, with the possibility of intrafractional adaptation and PTV margins as small as 3 mm (Kontaxis et al, 2017).

Prostate (Ghilezan et al. 2010; Pathmanathan et al. 2018; Vargas et al. 2010; Gill et al. 2014; McPartlin et al. 2016; Wu et al. 2008; Ghilezan et al. 2010; Schwartz et al. 2012): Much work has been done for both off-line and on-line adaptive strategies for prostate treatments following volumetric on-treatment imaging—either CBCT on in-room CT (Wu et al. 2008)—thereby trying to reduce margins and escalate dose for a better clinical outcome. Adaptive strategies are ideal for helping to reduce margins by systematically accounting for patient specific variations (Zou et al. 2018). Long-term clinical follow-up has shown encouraging results in clinical outcome and toxicity levels; average target-dose escalation has been nearly 10% through additional margin reduction using on-line adaptive protocols (Ghilezan et al. 2010). Because of the soft-tissue boundaries between the prostate and nearby OARs (rectum and bladder), MR offers improved anatomical visibility than, say, kV CBCT (Zou et al. 2018). Intrafractional motion during treatment delivery, proven by cine MRI, may mean that current margins are inadequate, if there is no compensation for the intrafractional motion (Vargas et al. 2010; Gill et al. 2014). With the appropriate protocol, PTV margins could be restricted to 3 mm or even less (McPartlin et al. 2016).

Bladder (Hafeez et al. 2017; Kainz et al. 2017; Foroudi et al. 2014; McNair et al. 2015; Murthy et al. 2016; Vestergaard et al. 2016; Gronborg et al. 2015; Kong et al. 2018a, 2018b): This is perhaps the clinical site that has had most studies for adaptive protocols, because of the constantly changing nature of the bladder through filling over time; inter- and intrafractional anatomical changes are likely. KV CBCT studies have featured heavily because of the superior soft-tissue image guidance compared with previous methods (Hafeez et al. 2017; Murthy et al. 2016); one would anticipate even better detail is possible under on-treatment MR guidance to allow margins to be reduced to about 5 mm (Vestergaard et al. 2016; Gronborg et al. 2015). Studies have often included the on-line plan-of-the day approach with a library of plans derived from isotropic and anisotropic margins grown around the planning PTV, repeat planning CT scans (at 0-, 15-, and 30-minute intervals) to form the library plans, patient-adapted single plans derived from the first five days of treatment, and even daily optimisation and replanning (Berkovic et al. 2017; Kong et al. 2018a, 2018b, Murthy et al. 2016; Hafeez et al. 2017; Vestergaard et al. 2016). Initial studies have reported good oncological outcomes with low acute and late toxicities for both standard and hypofractionated treatments (Hafeez et al. 2017; Murthy et al. 2016). The aspects for training for understanding the appropriate matching and on-treatment image analysis, and the decision-making processes for choosing the appropriate plan and applying it, should not be underestimated and is an on-going task for all radiographers/radiation therapists (McNair et al. 2015).

MR on-treatment imaging is likely to show clinical potential for clinical sites such as brain, head and neck, liver, pancreas, and prostate (Zou et al. 2018; Kupelian and Sonke 2014; Hunt et al. 2018; Pathmanathan et al. 2018; Choudhury et al. 2017) when coupled with adaptive algorithms specially written for MR imaging (Zou et al. 2018; Kontaxis et al. 2015; Ahunbay et al. 2016; Bol et al. 2013) and real-time anatomy tracking for certain clinical sites (Green et al. 2018; Olberg et al. 2018; Ramey et al. 2018). Figure 10.7 shows a developmental screen-shot of a typical MR guided adaptive workflow.

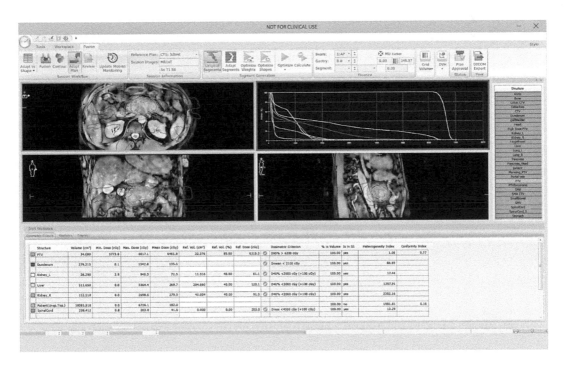

FIGURE 10.7   Developmental images for an MR-linac-based adaptive radiotherapy workflow.

Image courtesy of Elekta.

CHAPTER 11

Incident Reporting

11.1 INTRODUCTION

Radiotherapy is a major treatment option in the management of many types of cancer. It's one of the safest of treatments, saving, prolonging, and improving the quality of lives for millions of people around the world each year (Eaton et al. 2018; WHO 2008). Although our treatments have become more complex and more efficacious, there is still the risk of harm, personal tragedy, or even fatal injury to a very small percentage of patients (RCR 2008b; WHO 2008). There is a continual drive to reduce this risk, not just by the improvement of procedures and processes and by the use of more technology, but also by learning from mistakes and errors made and those that have nearly amounted to larger mistakes but were stopped beforehand (near misses). For this to be effective, it requires an open and honest no-blame reporting and review structure to be in place in radiotherapy departments around the world, feeding information through at local, national, and international levels (Pawlicki et al. 2017; RCR 2008b; Eaton et al. 2018; Findlay et al. 2016; Milosevic et al. 2016; Chang et al. 2014; IAEA 2016a; IAEA 2018; Ford and Evans 2018).

11.2 RATIONALE

There are a number of systems in place internationally that aim to achieve this, and they tend to be called incident learning systems (ILSs) (Eaton et al. 2018; Pawlicki et al. 2017; RCR 2008b). From these systems, we, as a multidisciplinary radiotherapy profession, can learn (especially from the near misses) how to prevent errors by disseminating information and learning from the reported mistakes so that practice can continue to develop in terms of both safety and clinical effectiveness (RCR 2008b). It is a practice that needs greater access and dissemination world-wide, as state-of-the-art radiotherapy itself needs to be more widely available around the globe (Knoos 2017); studies have already shown that ILSs can be effective at improving safety (Nyflot et al. 2015) and this is something needed globally (WHO 2008; RCR 2008b; Ortiz et al. 2009; Shafiq et al. 2009; Clark et al. 2010; IAEA 2000, 2014, 2016a, 2018; Knoos 2017; Pawlicki et al. 2017). As the work becomes more widely disseminated and practice develops internationally, it is important to try to rationalise the taxonomy used in describing—and also in reporting—radiotherapy incidents, adverse error events, near misses etc., (RCR 2008b; Malicki et al. 2014, 2018; Ford and Evans 2018; Pawlicki et al. 2017; Shafiq et al. 2009) based upon the risk to patient and potential (or actual) clinical consequence, so that information can be compared consistently internationally (RCR 2008b, Pawlicki et al. 2017, Ford and Evans 2018, Shafiq et al. 2009). There is still much work to be done in both these areas, which includes supporting independent initiatives between users to create an open forum, sharing valuable experiences, and obtaining freely available further practical education and training (e.g. i.treatsafely.org [Pawlicki et al. 2017]).

The purpose of this short section is not to discuss incidents and errors in radiotherapy in general, but to help the reader understand the ethos behind trying to make radiotherapy safer, both in the UK and globally; to be aware of typical reporting mechanisms and their rationale; and to consider on-treatment imaging within this process as a mechanism to help stop the larger errors (acting as the verification and safety net that it should be) while also considering what errors might be reportable through the use of on-treatment imaging. This is all part of general education needed for our clinical practice in the UK (HCPC 2013) and internationally (IAEA 2014).

11.3 UK EXPERIENCES

Within the UK, the guidance document *Towards Safer Radiotherapy* was published in 2008 in response to a drive for improving the safety of radiotherapy, and in conjunction with medical exposure legislation (IRMER 2017) that first came out in 2000 and the knowledge of some critical incidents that happened within the UK and indeed around the world (RCR 2008b; Knoos 2017; Ortiz et al. 2009). Its purpose was to analyse previous incidents (using well-established, practical approaches to incident analysis [Reason 1990, 2001]), to understand some of the contributory factors (particularly human centred), make recommendations for appropriate prerequisites for radiotherapy departments (such as having an externally accredited quality system) in order to improve safety, highlight particular practices that could help detect and prevent incidents, and to establish an open, nonpunitive, national reporting system for adverse errors, minor incidents, and near misses (RCR 2008b). Just some of the key points from the document are outlined below, as pertinent to safe on-treatment imaging and verification, but the reader is encouraged to consider these as part of an overall safety culture throughout a radiotherapy department and to examine this and other published literature (e.g. RCR 2008b; Pawlicki et al. 2017; Knoos 2017; Ford and Evans 2018; Shaifq et al. 2009; IAEA 2016a; WHO 2008; Chang et al. 2014; Montgomery et al. 2018 etc.).

- Factors contributing to adverse incidents can include lack of training, competency, and experience; fatigue and stress; poor design/documentation of procedures; over-reliance on technology and automation; lack of communication and teamwork; staffing and skills levels (continuous development and competencies); and working environments. All of these aspects have a very close and acute bearing on the safe and effective use of on-treatment imaging for geometric verification—and indeed in future techniques like adaptive radiotherapy.

- Classification of RT Errors into a radiation incident (RI), which might be: Reportable (Level 1); Nonreportable (Level 2); Minor (Level 3); a Near Miss (Level 4), or an Other nonconformance (Level 5). These are clearly organised into levels of severity and clinical significance. A pathway coding was also introduced to identify where in the normal radiotherapy patient and process pathway the incident occurred. Both were designed to help consistency of reporting and also of analysis, in the recommended open, national system (Eaton et al. 2018).

- Prerequisites for safety focused on areas such as (fully recorded) training and competency; multidisciplinary working and communication (between and within disciplines and teams); a quality system with procedures and work instructions fully documented; open questioning, regardless of position in the organisation; active and effective audit and review; and secure commissioning for technology and techniques. Once again, many of these aspects are critical to good and safe on-treatment imaging, and are discussed within this book.

- In terms of detection and prevention of incidents, patient ID and verification checks are particularly highlighted through initial planning and reference imaging checks (using final planned isocentre positions); laterality and relative shift checks; clear protocols for ensuring correct geometric set-up at the start (as a minimum) and throughout each course of radiotherapy (using good set-up instructions, checks, and on-treatment verification imaging); systematic review throughout treatments; and appropriate attention and investigation to concerns raised by staff at all levels.

- A national reporting system and mechanism (easily accessible, on-line) for consistent reporting of incidents and near misses, aiding analysis, national feedback, and learning for all departments.

For the national reporting system, it was recognised that the many minor incidents and near misses could help learning, and the possible prevention of major incidents, by identifying patterns of failure and predictable risks (Eaton et al. 2018). All UK incidents are voluntarily uploaded to the system (through Public Health England [PHE]), using standard classifications and coding for central analysis (Eaton et al. 2018; Findlay et al. 2016); feedback to the radiotherapy community and wider public is disseminated on a regular basis (e.g. PHE 2018a, 2018b, 2018c). This guidance and the reporting system have accompanied UK legislation; notably IRR17 and IRMER 2017 (as the latest statutes) (Eaton et al. 2018). Broadly speaking, any adverse incidents involving patient exposure (or now carers and comforters, too) are reported under the IRMER 2017 legislation (currently through the Care Quality Commission [CQC]); this is the case whether they are procedural (accidental and unintended exposures) or equipment-related. Any adverse events that are found to be equipment-related are generally reported under IRR17 (usually through the Health and Safety Executive [HSE]/Medical and Healthcare products Regulatory Agency [MHRA]).

11.4 UK LEGISLATION AND GUIDANCE

Until recently (IRMER 2017), reporting for incidents involving patients has focused on over-exposures (defined as those "much greater than intended" or MGTI). Recent guidance from the UK Department of Health (DoH 2017) has highlighted the new reporting of planning and verification imaging as part of the considerations for exposures MGTI – using a factor of 2.5 times the intended dose per imaging episode (DoH 2017; Eaton et al. 2018). Discussions are also being undertaken to consider guidance for under-exposures and positioning errors (geographic misses), also known to have serious clinical consequences (Eaton et al. 2018; Knoos 2017).

This new legislation and the guidance on MGTI has important consequences for monitoring and reporting doses involved for on-treatment verification imaging. As stressed in chapter 5, the dose is concomitant; and although the purpose is to verify and ensure that the (considerably higher) treatment dose is to be delivered as planned, with likely more serious clinical consequences if it is not, the doses cannot be overlooked. For this reason, over-exposures from imaging doses are to be considered and reported on as an incident, if necessary. The guidance (IRMER 2017; DoH 2017; Eaton et al. 2018) considers an episode as reportable if the imaging dose delivered is 2.5 times or greater than that intended. Here, an imaging episode is considered as what takes place for verifying a fraction, so it may entail multiple planar or volumetric images; the guideline factors are applicable to the total episode dose, not individual exposures within an episode. So if the imaging protocol specifies a number of repeat images can be acquired to establish the correct set-up, then these are not reportable as they are reasonable practice for optimizing the set-up. However, if the reasons are a failure to follow the protocol accordingly (for example through choosing a different imaging modality), and the total episode dose is 2.5 or more times the intended, then the incident is reportable (DoH 2017). Single exposures are unlikely to breach this limit; but (for example) choosing a volumetric imaging modality instead of a planar one could possibly do so. It is important that the intended be regarded as that required to verify the patient, not the maximum in a given protocol (Eaton et al. 2018). Scheduling is also something that should be noted, for instance, undertaking daily imaging rather than weekly or a situation where five unnecessary repeat exposures have been delivered for an individual patient over the course of treatment. These are to be flagged even if the 2.5 times dosage limit has not been breached to highlight potential systematic or clinical failures that could happen again if not suitably managed (DoH 2017; Eaton et al. 2018). In all cases, regular review and audit should be undertaken (as part of the quality system) to ensure that practice is optimised and to reduce any potential risk of adverse incidents.

Pertinent to on-treatment verification imaging is the subject of geographical miss. Clearly, total geographic miss has a significant serious effect on clinical effectiveness and was one of the reasons for introducing gross error determination within *On-target* (RCR 2008a). Partial misses are much harder to quantify in terms of clinical effect. The guidance (DoH 2017) highlights the use of locally defined error margins, which are recommended to be anatomically site-specific. These are not to be confused with planning margins or imaging tolerances for geometrical set-up. Instead, it is more like the concepts of gross errors, as defined in *On-target* (RCR 2008a); although achievable accuracy has improved enormously in the 10 years since *On-target* was published, values like 1 cm are likely not appropriate for some clinical sites (Eaton et al. 2018). There are no specifics within the MGTI guidance, but one would anticipate that any local error margin (LEM) chosen should be from a clinical evaluation of the significance of such set-up errors (DoH 2017; Eaton et al. 2018).

Within the UK, PHE publish regular reports, from analysis of all incidents (including reportable ones and near misses) uploaded nationally by radiotherapy departments onto their system, as a newsletter giving a general overview and information about current initiatives/good practice/events etc., (PHE 2018a) and a fuller analysis of the voluntarily submitted incidents from that quarter or year (PHE 2018b). Recently, they published a full 10-year report on clinical site visits (CSVs) as part of their task of analyzing and promulgating learning across the radiotherapy community (PHE 2018c), and in ensuring that timely and effective analysis of events, coupled with appropriate feedback, is given and that staff know their comments and notes are acted upon and valued (Pawlicki et al. 2017). Within this latest document are highlights for on-treatment verification imaging. Over the 10-year period, there were 20 CSV reports that discussed imaging protocols with both technical and training recommendations regarding poor image quality and staff training with respect to local imaging tolerances (PHE 2018c). National guidance was recommended for such cases (RCR 2008a; NRIG 2012). Good practice was also highlighted, for instance, where staff had access to flow charts within imaging protocols, with improvements in actual protocols being focused on in only five of the 20 reports after national clinical support was published (PHE 2018c).

In the latest annual report from the CQC (CQC 2018) about the radiotherapy notifications they received in 2017/18 (a decrease from the previous year), the majority of reports were in the submodality of planning/verification imaging. Some of the key themes emerging about on-treatment imaging mentioned that departments had begun to understand well guidance regarding notifications in this area, and that occasionally verification imaging identified "queuing" errors where the plan loaded for treatment was not intended for that patient undergoing imaging and treatment (CQC 2018). This highlights an important aspect for the use of on-treatment verification imaging: Its ability to help prevent more severe adverse incidents (e.g. delivering entirely the wrong plan to a patient, laterality problems, or even treating the incorrect patient) by taking set-up images early-on in the course of treatment delivery (as has been identified by, for example, Walker et al. 2015; IAEA 2016; WHO 2008). In the case of queuing errors, simple procedural changes can help lower this risk of an adverse incident. The notifications have highlighted technical errors associated with on-treatment verification imaging (e.g. miscommunication between the treatment delivery unit and the oncology management mystem [OMS] resulting in the scans being attached to the wrong patient). In some instances, depending upon the technology and processes within a department, it highlights human error as the cause in uploading the wrong reference data for a patient (CQC 2018).

11.5 INTERNATIONAL PERSPECTIVES

Incident learning (or reporting) systems have been established internationally in many countries (Eaton et al. 2018), but much more work is still required to further push the safety envelope for the benefit of our patients worldwide, especially those in developing countries (Knoos 2017). International reviews of these ILSs can be found in papers by Shafiq et al. (2009), Pawlicki et al. (2017), and Ford and Evans (2018). Examples include:

- North America and Canada (e.g. the PSO, RO-ILS and NSIR-RT systems): See Ford and Evans (2018), Hoopes et al. (2015), Walker et al. (2015), Montgomery et al. (2018), and Milosevic et al. (2016)

- Europe (e.g. the ROSEIS system and ACCIRAD project): See Coffey et al. (2009), Cunningham et al. (2010), Pawlicki et al (2017), Malicki et al. (2014), and Malicki et al. (2018)

- Australia: See Chang et al. (2014)

- Internationally through the IAEA (e.g. the SAFRON Project): See Pawlicki et al. (2017), Eaton et al. (2018), Knoos (2017), and IAEA (2000, 2016a, 2018)

But for these ILSs to be effective, there must be national support and sustainability (Pawlicki et al. 2017), especially through appropriate resources for the electronic systems and software (with on-going support) for on-line systems, and personnel for analysis, report generation, and site visits etc. Within the UK, the work of PHE and CQC are supported as government agencies. Within Europe, ROSEIS is professionally supported by ESTRO and ACCIRAD was a project initiated by the European Community. Internationally, SAFRON is held within a division of the IAEA, and the US systems (PSO and RO-ILS) are supported by the Center for Assessment of Radiological Sciences (CARS) and the American Society for Radiation Oncology (ASTRO)/American Association of Physicists in Medicine (AAPM) respectively (Eaton et al. 2018).

Within all the work discussed above, a common factor is risk; the risks involved to patients if adverse incidents occur that can be fatal in certain circumstances (Knoos 2017). Reporting and notifying on these adverse incidents, accidental and unintended exposures, near misses, and nonconformances is all part of the strategy of trying to minimise risks locally, nationally, and internationally. On the international front, this is an approach put forward by the WHO and IAEA (WHO 2008; IAEA 2016a). The ideas put forward suggest ways of screening for events that might result in an adverse incident, with a view to reprioritise efforts where risk might be greatest. It is based on a forward-looking evaluation and consideration of safety measures put in place and the potential consequences of adverse events. Barriers to reduce the overall risk (based upon probability and severity of events) are proposed in terms of robustness—Type 1: Interlocks; Type 2: Alarms; Type 3: Procedures undertaken by different people; and Type 4: Procedures undertaken by a single person (IAEA 2016). With an international perspective, they highlight incidents in both well-established, high-income countries as well as those with lower incomes and less developed radiotherapy profiles; the methods can be used in all contexts. In terms of on-treatment verification imaging (described as portal imaging in the documents), focus is made on the incorrect or nonuse of such verification methods—the latter quite likely in those centres that have less well-developed profiles. It is seen as an important barrier (type 3 or 4) (IAEA 2016) that can be effective against adverse incidents before first treatment, weekly, or more frequently still in more developed centres with greater resources. The interpretation, analysis, and decision-making might be solely conducted by therapeutic radiographers/RTTs (e.g. in the UK) or by physicians at first fraction—all to ensure correct set-up—but in doing so, simultaneously helps to minimise the risk of an incorrect set-up or even incorrect patient (IAEA 2016a; Walker et al. 2015; WHO 2008; Shafiq et al. 2009). However, the barrier can only be effective if the personnel using it are effective at using, interpreting, analyzing, and actioning the clinical imaging protocols; training and maintained competencies are important perspectives, which have been discussed in chapter 4 and will be again in chapter 12, and highlighted by a number of international documents (WHO 2008; IAEA 2016a; HCPC 2013; Shafiq et al. 2009; Ford and Evans 2018; Pawlicki et al. 2017). Appropriate competency and training is required of the full multidisciplinary team (clinicians, physicists, radiographers) so that on-treatment verification imaging can be an effective barrier; inappropriate training/competency could result in incorrect patient positioning, unnecessary adjustment, weak imaging protocols, or an unnecessary concomitant dose (hence the changes for IRMER 2017 and MGTI definitions [DoH 2017; Eaton et al. 2018]). Since a multidisciplinary approach is vital, communication and teamwork (both within and between disciplines) is also vital (RCR 2008a; Ford and Evans 2018; Shafiq et al. 2009).

One might bring all of these aspects together, as some authors have already done (e.g. Pawlicki et al 2017) to summarise the overall steps required for successful design and implementation of an ILS, beginning with considering risk in the first place:

- Consider risk analysis of radiotherapy patient pathway, processes, procedures, and technology

- Ensure appropriate commissioning is undertaken (for both on-treatment verification imaging equipment and the protocols to be used)

- Establish the ethos, strategy and support for the ILS, especially in terms of resources, before adverse incidents occur

- Ensure suitable and adequate training and competency, for equipment, processes and the open, no-blame reporting culture internally and externally

- Consider an immediate response after any adverse event

- Utilise internal and external reporting mechanisms as appropriate

- Prepare for an appropriate analysis of the incident

- Perform full event analysis, depth, and scope as deemed necessary

- Sharing open and honest analysis with external site visitors (from national bodies) as necessary

- Implement changes, workarounds, and procedural changes as necessary; consult with the wider community and the output from the ILS

- Share incident information and lessons learned through the ILS and the radiotherapy community as a whole.

CHAPTER 12

Protocol Development and Training

12.1 PROTOCOLS—THE RATIONALE

Protocol development and training are integral parts of commissioning a new linac or treatment machine with its associated on-treatment imaging and geometric verification capabilities or, indeed, when a new verification technique is introduced on existing equipment. Also, when one is responding to quality assurance, audit, or protocol review, development and training are important, especially when changes are required in light of errors occurring, aspects of the current protocol are difficult to execute or are not being executed, or if new evidence has come in from the literature or other centres. Protocols are an explicit requirement of national standards and quality programmes in cancer in the UK (e.g. NCAT 2013; NHS 2018); they are mandatory for geometrically verifying the accuracy of treatment delivery and implied under legislation (IRMER 2017) as part of a QA programme (see chapter 5), ensuring all exposures involved in a patient's treatment are authorised as justified and, in a clinical sense, in terms of optimisation. In all cases, protocols must be developed and integrated into the quality management system (see chapter 9), and staff appropriately trained in order to gain optimal clinical use and the best outcomes for patients.

As mentioned in chapter 10, one of the best templates for protocol development, which covers numerous aspects comprehensively perhaps because it applies to a new, initially unfamiliar treatment technology, are a number of papers that describe the clinical development and implementation of tomotherapy. Notable are publications by Burnet et al. (2010); Dean et al. (2013) and Bates et al. (2013). These are excellent examples of protocol development for implementing new technology as a whole and on-treatment verification imaging protocols in particular (Burnet et al 2010; Bates et al. 2013). They are ideal in that their approach is without foreknowledge or experience. Quality assurance (both technical and clinical) and audit are vital aspects for monitoring their suitability for clinical use. *On Target* (RCR 2008a) and NRIG (2012) as national guidance documents, examined the basis of on-treatment imaging and verification protocols and provided generic examples especially for traditional linac-based practice and for the direction of travel for developing practice, especially toward 4-D adaptive RT (NRIG 2012). RCR professional guidance on IRMER (RCR 2008c) and *Towards Safer Radiotherapy* (RCR 2008b) allude to safety aspects especially as far as concomitant dose are concerned (see chapter 5) and to the safe delivery of radiotherapy in general, which naturally must include on-treatment verification imaging. Ensuring the dose is delivered as planned, quantitatively and geometrically, is paramount for all treatment techniques and modalities, especially as far as highly conformal methods are concerned like IMRT/VMAT and charged particle therapy.

12.2 DEVELOPING PROTOCOLS FOR ON-TREATMENT VERIFICATION IMAGING

NRIG (2012) and *On-target* (RCR 2008a) give the basic structure required of protocols for on-treatment verification imaging; the key points are outlined here and are equally applicable for C-arm based linacs and other advanced technologies like Cyberknife, TomoTherapy, Halcyon and MR-guided radiotherapy. The balance for the newer technologies might be more toward on-line protocols for adjustment of set-up correction at the start of each fraction (like CBCT) or indeed throughout the fraction (i.e. intrafractionally).

In all cases, site specific protocols are ideal; ones that can be tailored to the nuances of the clinical site and the best treatment delivery method for that site. For example, prone/supine patient set-up; the immobilisation equipment used; the type of treatment technique used (conventional and conformal or IMRT/VMAT); whether the intent is radical or palliative; and the general fitness of the patient themselves. The methodology of set-up error correction still considers the three main types of error as **gross, systematic** and **random**.

All protocols should have some element of gross set-up error detection; a clear definition of what constitutes a gross error (what are the tolerances involved) and a clear understanding of action levels. For example, as simple an understanding that a 10 mm gross-error tolerance means that set-up errors of 10 mm and above will be actioned upon immediately. But perhaps a part of the technique protocol as a whole (in advance of the on-treatment imaging protocol) must come from aspects of good clinical training and common sense— so that the first tools of verification are not images (and therefore a likely concomitant dose delivered to the patient), but a thought process of what to expect and anticipate during set-up. Perhaps a key part is simply knowing that if things look wrong, then that is the time to stop, think, and re-evaluate, rather than needing an image or series of images to do so.

Protocols are ideally still divided into **off-line** and **on-line**; even the advanced technologies that are likely to correct set-up error (each day or continuously) fall into the on-line category here, but perhaps in a slightly more sophisticated or less obvious manner. Flow diagrams for generic and site-specific application are ideal for both designing the protocol and its routine use. Excellent examples are given in *On-target* (RCR 2008a) and NRIG (2012). Bates et al. (2013) outlines well the development, implementation, and refinement of their imaging protocols for TomoTherapy in a research setting and one that primarily is based on daily, on-line procedures. They highlight image quality, dose, and practical implications for resources (human and set-up time) in a busy clinic from data accrued over five years through more than 18,500 images for clinical sites like prostate, head and necks, and CNS.

The intention is always to assess the acceptability of patient set-up for the daily treatment, trying to eliminate both systematic and random errors, which always require multiple images in order to gain a reasonable assessment of these. Recalling our definitions in chapter 1:

- **Systematic Errors** represent a geometric shift in the high-dose volume and dose distribution (modelled as a rigid body) with respect to the target volumes; it is likely a consistent error, similar in direction and magnitude. It has a number of possible causes; differences in equipment (through the patient pathway from pretreatment imaging and scanning through to treatment); systematic anatomical changes since the planning scan; and human errors in isocentre placement or reference moves etc. Systematic errors tend to have the largest clinical effect on generally decreasing dose-to-target and increasing dose-to-OARs (RCR 2008a).

- **Random Errors** represent a blurring out of the dose around the target from set-up errors in multiple directions, caused generally by internal anatomical changes (daily between fractions and/or intrafractionally); changes in target volume positioning and shape; machine tolerances in mechanical and beam movements; immobilisation differences, errors or malfunctions; and human errors in set-up.

Offline protocols (see chapter 6) were developed because the systematic error has the greater clinical effect and in attempts to balance the dose and resource implications (concomitant dose to patient; time of patient set-up with impact on workloads and accuracy of treatment itself); so sufficient daily images were acquired to give a good estimate and elimination of systematic error without daily imaging and continual set-up correction. Examples include SAL and NAL protocols (Bel et al. 1993; de Boer et al. 2005; de Boer and Heijmen 2007); the NAL being the most popular and least resource-intensive, perhaps requiring three to five images for systematic error estimation and then with/without weekly confirmation imaging (with potential updates on set-up correction within the treatment [RCR 2008a; de Boer and Heijmen 2007]). The potential changes in set-up intrafractionally are still of issue, but their effects are less significant if the set-up and treatment delivery are quick and efficient (e.g. using VMAT).

Online protocols attempt to check and correct set-up errors prior to delivering the daily fraction, thus eliminating both systematic and random components simultaneously. For minimal resource implication (especially in terms of increased set-up time and concomitant dose), perhaps fast 2-D imaging (e.g. kV-kV or kV-MV orthogonal image pairs), with automatic image registration/matching and gantry and couch-correction from outside the room is an ideal technique for many clinical sites. But one must be aware of the clinical arguments of 2-D vs. 3-D imaging and overall risk vs benefit. A technique involving manual matching places added pressures on the radiographers/radiation therapists for decision making in an often time-pressured environment. Thresholds for action (to effect couch correction) against correction of all set-up errors irrespective of magnitude must be considered together with concomitant dose and adequately trained staff for daily on-line protocols. A compromise is a protocol that is not performed for every fraction, but is limited in number (similar to the NAL offline protocol described above); estimation and elimination of the systematic error is still needed, however, with possibly considering on-treatment imaging at the end of each fraction on a number of occasions to consider intrafractional changes and/or residual errors.

For all types of protocol, there is an onus on commissioning and frequent review/audit of practice and results as well as feedback into maintaining/modifying the protocol for maximum clinical quality. These could be enacted easily within the principles of QA programmes and the QMS. Key points (as part of commissioning and routine QC) could include:

- Check and evaluation of matching techniques (manual and automatic) in terms of their robustness, efficiency, and training for use; consistent use of surrogates and training/evaluation for soft-tissue matching techniques

- Checks on remote couch corrections, especially for small magnitude and directions of corrections. Special attention should be paid to mechanical hysteresis in movements and to the veracity of corrections if dataflow is involved before couch correction for online work (e.g. from linac => OMS for analysis => linac for couch correction).

- Evaluation of concomitant dose, effects on image quality, and effectiveness for set-up error measurement and correction

- Dose limits maintained within the protocols by quantitative estimation and/or limits to number and types of on-treatment imaging; avoidance of critical structures for open-field imaging (especially 2-D MV imaging), unnecessary image volumes; alignment with legislation and national guidance

- Audit of actual clinical practice compared with protocols by examining the need for repeat imaging, steps being missed, new steps introduced, incidents and near misses, repeat imaging or wrong imaging, and timing of imaging and time studies of acquisition, online and offline analysis etc.

Data and information from review should be used for margin calculations (RCR 2008a; NRIG 2012) and for refinement of protocols to obtain the best balance of safe, secure, and effective clinical practice for the patient and resources within a department. Excellent clinical examples of protocols are given in *On-target* (RCR 2008a; NRIG 2012); and in other publications such as Burnet et al. (2010); Bates et al. (2013); Dieterich et al. (2016b); de Los Santos et al. (2013); RCR (2008c); IRMER (2017); and RCR (2008c). It should be noted that protocols are unlikely to be exactly the same from one department (cancer Centre) to another; there are always differences in resources, equipment, clinical directives, and experiences. However, they are likely to have the same generic basis or be very similar across a network of centres, for example, between a main site and its satellite centre(s), and best operated under the same standards/practices of QA programmes and the same QMS. But even here, local nuances may be needed, most notably in terms of available human resource. As a simple, single example, the availability of experienced staff for an online image that might be on the edge of tolerance or require an adaptive rather than geometric approach, this might be more readily available in a larger centre with a pool of resources than a smaller one (or satellite centre); although with modern networking power (and the same OMS across the centres, with the availability of remote viewing and teleconferencing) the availability of expertise and advice is now much improved. But the process and action required may therefore be different and written specifically into the on-treatment imaging protocol.

As a final point, the on-treatment imaging protocols must always be considered a part of the full clinical protocol for that particular clinical or disease site; it is a part of the whole radiotherapy process for a patient—an integral and vital part—but one that must not stand alone in isolation. One most obvious reason is the interplay with treatment planning and simulation for the reference images for on-treatment imaging, the feedback to treatment planning for CTV-PTV margin calculation and refinement, or the feedback to clinical review for toxicity and treatment reactions etc.

12.3 TRAINING—THE RATIONALE

Appropriate, adequate, and timely training must be emphasised for all aspects of radiotherapy, not just for on-treatment verification imaging. There is an emphasis on training and indeed a requirement through a number of platforms, e.g.:

- Under national guidance (RCR 2008a; NRIG 2012; RCR 2008b; RCR 2008c)

- Through national standards (NCAT 2013; NHS 2018) and government agencies (PHE 2018a, 2018b)

- A mandatory requirement under legislation (IRMER 2017; IRR17)

- Through international guidance (Ortiz et al. 2009; IAEA 2014, 2016; RCR 2008b)

- Through publication from centres and national bodies highlighting good practice (Burnet et al. 2010; Dean et al. 2013; Bates et al. 2013; Zou et al. 2018; de Los Santos et al. 2013; Dieterich et al. 2016a, 2016b; AAPM 2018; IPEM 2018; RCR 2018; SOR 2018; BIR 2018; ASTRO 2018; AAPM 2018; ESTRO 2018)

- Through publications and systems highlighting errors, incidents, and near misses (Pawlicki et al. 2017; Malicki et al. 2018; Ford and Evans 2018; RCR 2008b; Johnston 2006, 2015; Bogdanich 2010)

It is something engrained into the core curricula for all professionals associated with radiotherapy (e.g. IAEA 2014; Potter et al. 2012; Eriksen et al. 2012; EUMS 2013; IAEA 2013b; RCR 2016; RANZCR 2018; NSHCS 2014; ACPSEM 2013; SCoR 2013; HCPC 2013;

Burmeister et al, 2016; Paliwal et al. 2009; Kirby 2015; Bridge et al. 2017) as part of education and training preregistration and clinical practice; but also for clinical competency and continued professional development throughout the health care professional's working life (Kirby et al. 2014; BIR 2014; BIR 2016; BIR 2018). On-treatment imaging training is a part of this whole, for developing, maintaining, and improving standards and care for patients.

For on-treatment verification imaging, again national documents like *On-target* (RCR 2008a) and NRIG (2012) outline the requirements nicely, but with image guidance being available and used on all equipment now in the UK, the emphasis of extending skills and training for geometric verification is now much less about a specific team who, for example, may be involved with CBCT for the first time on one machine in a department, to now being the requirement for all staff on every machine; such is the extended use of on-treatment imaging now for both the more straightforward to the most complex of techniques. Where advanced, non-C-arm technologies are concerned, on-treatment verification imaging is often an integral system and therefore a necessary part of the overall training of the radiographer, physicist, clinician, and engineer.

For the core service for image guidance and on-treatment imaging, a multidisciplinary approach is necessary and the best practice. This means continuous training for maintaining knowledge, understanding, and skills for what is a fast-changing practice through technological and technique evolution. Secure training in a professional's own discipline is needed, but also an appreciation of the skills and experience needed for all the other professionals involved. Competencies must be regularly reviewed and maintained; with preregistration, undergraduate, and postgraduate training updated to reflect the continually changing landscape.

Core skills must include a knowledge, understanding, and appreciation of various aspects of the service—depth being established in perhaps a core leadership team, but extended to all staff, since all are involved in image guidance and many in on-treatment imaging for geometric verification. A local team (of clinicians, radiographers, physicists, and engineers) is required, with expertise spread across the team to cover the appropriate depth, but dependent upon local practice, implementation, resources, and technology in clinical use. The skills should include, but not be restricted to:

- Understanding of equipment and technique design and implementation

- Site-specific protocol development and local knowledge

- Off-protocol decision making and requirements

- Image acquisition and review skills

- Concomitant doses for different technologies and techniques; dose limits for on-treatment imaging and justification/authorisation beyond

- Assessment for repeat imaging, replanning, adaptive therapy

- Audit and review through the QMS and QA programmes

- Applications and cascade training; local training

- Technical maintenance and service; technical QA programmes and requirements

- Clinical QA programmes and requirements

- Legislative, standards and guidance requirements for techniques and dose requirements

- Appreciation of the wider evidence base, technology, and techniques

- Maintaining quality through QA programmes and the QMS

One might identify discipline-specific requirements for training, which could include, but again are not confined to those highlighted in Table 12.1. It should be a requirement of all disciplines to have awareness of the need and benefits from regular review and audit of practice, including the risks and benefits associated with on-treatment imaging for the techniques and technologies used and an appreciation of incidents and near misses, their consequences and all persons' parts in minimizing their occurrence. There is naturally an understanding that roles and responsibilities vary between departments, cancer centres, regions, and countries.

As roles develop, specialist knowledge, understanding, and training may be required for the current team leaders, but also for the leaders of the future in all disciplines. This might include:

- Advanced training courses and postgraduate modules, for example, through professional bodes like SCoR, RCR, IPEM, BIR, ESTRO, AAPM, ASTRO, RANZCR, and ACPSEM (referenced earlier in this chapter); through refresher courses at national and annual meetings; and postgraduate modules and CPD

TABLE 12.1   Example Discipline-Specific Training Requirements for On-treatment Verification Imaging

Discipline	Training Requirements (e.g.)
Clinician/Physician (Clinical oncologist/radiation oncologist)	Image interpretation; set-up errors, anatomical changes and their clinical effects on dose distributions, target, and OAR doses
Radiographer (Radiation therapist); Dosimetrist	Image acquisition, registration/matching, interpretation; protocol design, implementation, review; decision making; concomitant dose; set-up error reduction; incident reduction and investigation; quality assurance
Physicist	Imaging hardware, acquisition, registration/matching algorithms and methods, interpretation; image quality, dose, and system QA; networking and image/data storage; set-up error reduction; incident reduction and investigation
Engineer (Maintenance technician/service engineers)	Equipment calibration, QA, servicing and maintenance; technical training and operation; technical and clinical functionality; mechanical and systems tolerances; software and hardware interfaces including networking

- Teaching and training methods and experience; competency framework training and design

- Research and development

- Experience in commissioning and implementation of new techniques, protocols and technologies

- Systems and protocol development; set-up error reduction; minimisation of incidents; their investigation and reporting; margins and feedback to planning processes

- Soft-tissue matching and adaptive procedures; clinical decision making regarding TV coverage and OAR sparing

- Project and management team skills and leadership; team building

- Anatomical changes and their clinical consequences; plan modification and replanning; automatic and fast planning techniques and development

- AI and machine learning for clinical processes

- Change management following review and audit

- Managing and leading in assuring quality

- Research and involvement in external clinical trials

It should be emphasised that research and development is not a subject area confined to a few members of each discipline; it is the necessary clinical practice of all professionals to have a desire to question, progress, and develop the service for the benefit of patients. This is the fundamental foundation of research and, therefore, the preview of all.

CHAPTER 13

Commissioning of New On-treatment Imaging and Techniques: Integration Into the Oncology Management System

13.1 INTRODUCTION

This chapter examines two separate, but not unrelated components of on-treatment imaging for geometric verification. The first is the commissioning of such equipment, techniques, and the processes involved (i.e. the clinical protocols). The second examines the oncology management system (OMS), often termed internationally as the record and verify (RVS) system, its expanding role in radiotherapy, and how on-treatment imaging is now fully integrated into its functionality. Although separate subjects here, their functions are not unrelated; commissioning is required for the OMS as much as it is needed for the equipment and technologies involved in on-treatment imaging and radiotherapy itself.

13.2 COMMISSIONING

Commissioning is a vital part of the introduction of any new piece of equipment or process into radiotherapy for first clinical use. Most may be familiar with the commissioning of the linac, as the detailed period of tests and measurements (often taking weeks or even months) that may be undertaken on a new machine prior to first clinical use. It is vital in that the data collected sets the baseline for clinical use for the potential lifetime of the equipment or the technique being developed; it also holds an associated risk in that any critical component or data incorrectly measured or process established at this stage could go unnoticed and could pose a clinical risk to patients and/or staff. Notable incidents

TABLE 13.1 Descriptions of Typical Acceptance and Commissioning Processes Involved for Radiotherapy Equipment

Acceptance Testing	Defined by the manufacturer for demonstrating compliance with IEC standards and any particular customer requirements at the time of equipment specification and purchase
	Conducted by customer/installation engineer immediately after installation
Commissioning	Detailed measurements, testing, and data collection following installation and prior to clinical use
	Can be multidisciplinary, particularly for software and systems that (e.g.) clinicians and radiographers would use
	Sets baseline values for future quality-control checks; produces data charts; collects data for (e.g.) treatment planning system
	Includes training, writing work instructions and procedures, training guides etc.
	Includes all dosimetric checks, including definitive calibration when considering linac commissioning

in machine calibration, TPS commissioning, and treatment delivery have been related to procedures undertaken and errors introduced/missed at the commissioning stage for new equipment or techniques (Ortiz et al. 2009; Johnston 2006).

For radiotherapy equipment, the commissioning phase will follow the formal acceptance process; a stage that has both legal and financial implications for the manufacturer and the customer (the hospital or cancer centre). Its aim is to "….ensure that the equipment, departmental procedures and data are adequate to support delivery of precise and safe treatments to patients" (Kirby et al. 2006b). Useful descriptions of typical acceptance and commissioning processes are given in Table 13.1.

The commissioning tests and checks establish the baseline for clinical operation and routine quality assurance (see chapter 9) and define the expected standard values for quality-control checks undertaken routinely for daily, weekly, monthly, and annual checks (Patel et al. 2018). By far, the majority of these standards are technical parameters, but we must not forget that commissioning is equally importantly and required for the establishment and implementation of new clinical techniques, too, including those used for on-treatment imaging, such as different clinical sites. Setting up and creating the protocols, procedures, and work instructions for clinical use of the equipment for different anatomical treatment sites and the training of staff (as we saw in chapters 4 and 12) are equally important commissioning tasks as are establishing the technical abilities and parameters for the equipment. As identified in chapter 9, incorrect protocols could affect the efficacy of treatment of many patients on an individual machine, or indeed across a whole department, with clinical consequences ranging from geographical miss of target volumes to increased toxicity/reduced local control to unnecessary concomitant exposures for the patient (and therefore increased risks of secondary malignancies).

In considering solely on-treatment verification imaging, commissioning could therefore include:

- The on-treatment imaging and geometric verification aspects of a whole new treatment delivery unit (Kirby et al. 2006b; Patel et al. 2018); one perhaps that is completely new to the department as the first of its kind and perhaps the only one in the department (Burnet et al. 2010). The process is equally important if it is a similar, matched machine to others in the department, or a completely new type of treatment delivery unit (such as were discussed in chapter 10); the timescales for these two examples are likely to be completely different, however.

- The addition of new on-treatment imaging equipment onto an existing treatment machine or into an existing treatment room. For instance, in adding kV- or MV-based 3-D CBCT volumetric imaging onto a linac that previously used only 2-D MV imaging, or adding in-room stereoscopic kV imaging into a treatment room with an existing linac.

- The set-up of a new clinical technique using on-treatment imaging equipment, after the treatment machine has been in clinical use (perhaps with a minimal, or even no use of the on-treatment imaging capabilities). For example, when moving from conformal therapy to VMAT; or moving from 2-D imaging based protocols with fiducial markers to a 3-D or 4-D based volumetric technique.

- Changes in existing clinical imaging technique, for example a move from off-line imaging and a NAL-based protocol to an on-line imaging protocol with daily set-up correction.

In many of the cases above, the commissioning may involve acquiring and establishing the baselines for the more technical aspects of the equipment and techniques. For example:

- Baselines for the quality assurance programme (see chapter 9), which could include image quality standards (e.g. noise, spatial and contrast resolution), dose standards (which could affect frequency of imaging or type of imaging protocol to be undertaken), and imaging and treatment isocentre checks etc. Here, image quality phantoms could be used for both 2-D and 3-D imaging, specific dosimetric equipment for exposure measurements, or ball-bearing phantoms for checking and monitoring coincidence of isocentres and calibrating flexmaps.

- Tests of scaling, distance, orientation etc. for 2-D, 3-D, and 4-D imaging methods; image integrity under data transfer, for example from CT scanner => TPS => OMS => linac => OMS. Checks could include specific 2-D and 3-D phantoms, anthropomorphic phantoms, or DICOM based image libraries.

- Tests and scrutiny of the veracity of image registration and matching routines, resultant remote couch movements, couch readouts etc. Here, checks might make use of both digital image libraries (libraries of images with predefined scaling and orientation and with known shifts and rotations from a reference image) and/or physical phantoms that can be accurately adjusted in 2-D/3-D to mimic translational and rotational set-up errors from a known reference image.

- Checks made when equipment or software is newly installed; also where there are significant changes or upgrades of software on existing equipment.

It would also include developing and formulating the protocols for clinical use for different anatomical sites (RCR 2008a; Burnet et al. 2010, NRIG 2012). These might be based upon current techniques within the department or the results of audit, evidence review, and/or the desire to develop new techniques (where techniques could be both the clinical delivery of treatment and/or the imaging technique itself). The resulting imaging protocols might also be dependent upon technical imaging capability (e.g. image quality, dose, 2-D, 3-D, or 4-D techniques) or the technology itself (e.g. if imaging panels can be deployed from outside the room or not). All these aspects are the basis for the imaging protocols described in chapters 4, 6, 8, and 12.

Ideally, any type of commissioning should involve a carefully developed plan and schedule for testing, data acquisition, analysis, protocol development, training and implementation. It should be undertaken by a multidisciplinary team chosen for their expertise in particular areas and related to the tasks for clinical use, but also engaging with fresh ideas and minds in order to develop teams for the future. For example, this might mean physicists and engineers for some of the more technical aspects; radiographers, radiation therapists, and clinical staff, for the development and issue of imaging protocols; and a multidisciplinary team for the whole process for various clinical sites and routine use. But these suggestions are open to flexibility and also, practically, to the available resources—both human and technical.

As mentioned above, and detailed in chapters 4 and 12, training is an integral and vital part of the commissioning process. There is little point in establishing an elegant imaging protocol for a particular clinical site, or setting up a new imaging panel with the best image quality available, if staff are not trained in the use of that protocol or in the best way to routinely test and maintain the image quality. The individuals most likely to be adversely affected as a result of this are the patients. This could mean their treatment is not as accurate and precise as we might be able to achieve; or, in the worst case scenario, patients are actually treated improperly as a result (geometrically or dosimetrically re concomitant dose). As an integral part of commissioning, training could include:

- Applications training from the manufacturers; the foundations to how the equipment is used and its software interfaces. These might be changes from existing software or radically new interfaces.

- Design and use of departmental work instructions/procedures written and tested as part of the quality system for using the equipment (see chapter 9). Cascade training could be used to train other staff accordingly, for those who will use the equipment the most in a clinical manner.

- Review of current protocols, literature, and site visits, from which one can establish site-specific clinical protocols for on-treatment imaging. Training for staff might perhaps include anthropomorphic or similar phantoms alongside imaging and clinical protocols that have been written in end-to-end tests for the whole radiotherapy process, including all the aspects of pretreatment and on-treatment imaging.

Although described here as commissioning, another term often used more widely is clinical implementation. To a certain degree, our description of commissioning includes all the aspects of clinical implementation as well as the technical aspects of the checks and baselines for quality assurance of the equipment itself. Both are vitally important for assuring quality of the treatment delivered for patients—the very heart of the purpose of on-treatment imaging for geometric verification. Within the literature, perhaps a seminal work on commissioning is that published by Burnet et al. 2010. Although pertaining to a non-C-arm-based linac technology, it includes the practical elements of both the technical and clinical aspects in the goal of clinical implementation, and has already been introduced in chapter 12.

The literature is a mine of information for key writings on commissioning and clinical implementation for on-treatment imaging equipment and techniques. A sample of key papers are outlined here, in chronological order, in Table 13.2.

13.3 INTEGRATION INTO THE ONCOLOGY MANAGEMENT SYSTEM

When on-treatment imaging for geometric verification was first implemented in the form of electronic portal imaging (from first- and second-generation MV EPIDS to the latest kV/MV CBCT and nonionising radiation based methods [see chapter 7]), the computer workstations for acquisition, display, and analysis were often stand-alone systems, usually solely on an individual treatment machine (Langmack 2001; Antonuk 2002; Herman 2005; Kirby and Glendinning 2006; IAEA 2007; van Herk 2007; Evans 2008). As the technology and computer networking technology developed, these systems became interlinked; for example, at first by means of a common image database on a shared server for patient data storage and access from multiple linacs through to the present day where it is an integral part of a fully networked department by means of the oncology management system, also termed the oncology information system, (Elekta 2018a, Varian 2018, RaySearch 2018) which itself developed from the principles of record and verify for patient treatments (Kirby et al. 2006a; Kirby and Davies 2010; Symonds et al. 2012; Walter and Miller 2012; IAEA 2013a; Shakeshaft et al. 2014). Patient data is now centralised (often utilising a single database for patient preparation, pretreatment imaging, simulation, computerised planning, on-treatment imaging, and treatment delivery) storing both parameters and images using DICOM-RT established formats, and networking them using internet-based transmission protocols (See Figure 13.1).

Its functionality has, therefore, grown dramatically over the years (Kirby and Davies 2010, Shakeshaft et al. 2014):

- First, from that of record and verify (RnV) on the linac alone, which was nonimage-based and used to verify the set-up of the actual treatment plan parameters and record delivered parameters, including monitor units (Kirby et al. 2006a, 2006b)

- Next, to a system that can integrate booking, scheduling, pretreatment imaging and simulation, computerised treatment planning, RnV, and image-based on-treatment geometric verification as illustrated in Table 13.3. This naturally made the database functionality

TABLE 13.2 Some Key Published Works for the Commissioning and Quality Assurance of Radiotherapy Equipment Pertinent to On-treatment Verification Imaging (in Chronological Order)

Year	Reference	Outline
1993	Lin et al. 1993 (AAPM TG 2)	Specification and acceptance testing of CT scanners
2000	Philips et al. 2000	Commissioning for image-guidance for SRS, conformal, and IMRT techniques
2005	Herman 2005 (AAPM TG 58)	Clinical use of electronic portal imaging—technical aspects, development of protocols, etc.
2006	Kirby et al. 2006a (IPEM 93)	Commissioning and quality assurance of a networked radiotherapy department
2006	Kirby et al. 2006b (IPEM 94)	General commissioning of linac and on-treatment imaging equipment
2007	Lehman et al. 2007	Commissioning CBCT on-treatment imaging equipment
2008	RCR 2008a (On-target)	Development of clinical protocols for on-treatment imaging for geometric verification
2009	Klein et al. 2009 (AAPM TG 142)	QA of medical accelerators
2009	Yin et al. 2009 (AAPM TG 104)	In-room kV imaging systems (CT-on-rails, stereoscopic in-room kV imaging, gantry-mounted kV CBCT, hybrid systems)
2010	Burnet et al. 2010	Commissioning of new clinical equipment and on-treatment imaging techniques (tomotherapy)
2010	Korreman et al. 2010	Practical and technical review of 3-D CT-based on-treatment imaging technologies
2011	Dieterich et al. 2011 (AAPM TG 135)	Commissioning and QA for robotic RT systems (Cyberknife), including end-to-end tests
2011	Molloy et al. 2011 (AAPM TG 154)	QA for ultrasound systems for external beam radiotherapy
2012	NRIG 2012	Development of clinical protocols for on-treatment imaging
2012	Bissonette et al. 2012 (AAPM TG 179)	QA for on-treatment imaging using CT-based technologies
2012	Willoughby et al. 2012 (AAPM TG 147)	QA for nonradiographic radiotherapy localisation and positioning systems
2012	D'Ambrosio et al. 2012	Continuous localization for radiotherapy using nonionising radiation technologies
2013	Arumugam et al. 2013	Phantom for assessing 2-D and 3-D registration algorithms and couch corrections
2013	IAEA 2013a	Acceptance, commissioning, and quality control of record and verify systems (RVSs)
2013	De Los Santos et al. 2013	Critical review of IGRT technologies
2014	Fontenot et al. 2014	Commissioning and QA for x-ray-based on-treatment imaging technologies
2016	IAEA 2016b	Accuracy requirements and uncertainties in radiotherapy
2017	Brock et al. 2017 (AAPM TG 132)	Image registration and fusion algorithms
2018	Patel et al. 2018 IPEM 81 (2nd edition)	QC in radiotherapy
2018	Ding et al. 2018 (AAPM TG 180)	Image guidance doses delivered during radiotherapy

FIGURE 13.1   A photograph of the typical control area for a modern linac showing how the control interface has changed from being solely record and verify for the treatment unit, to having full access to on-treatment verification imaging (acquisition and on-line analysis), booking and scheduling, treatment planning, etc.

Image courtesy of Elekta.

and size grow and change quite dramatically (Francis and Sipahi 2014; Elekta 2018b), but now maps every single part of the patient's clinical (or radiation) oncology and medical oncology pathway, including treatment planning (See Figure 13.2).

- Now, it can go beyond the standard C-arm-based linac technology to also integrate data networking, databasing, and recording appropriate for nonlinac and third-party systems, such as SXT, Gamma Knife, brachytherapy, proton-beam therapy, PACS, HIS, RIS, EMR, EPR, etc. (Francis and Sipahi 2014)

TABLE 13.3   Some Features, Specifications and Functionality of Modern OMSs (OISs) (Elekta 2018a, 2018b; Varian 2018)

• Patient Management; EMR/EPR functionality; diagnosis and test management; patient notes in consultations, review clinics, follow-up clinics
• Clinical Assessment: Response to treatment, review reactions, electronic prescribing, lab results
• Treatment Plan Management: Interface with different planning systems, different VSim software
• Treatment Delivery: DICOM interface protocols to different linacs and other RT systems, e.g. TomoTherapy, Cyberknife, Gamma Knife etc.
• Scheduling and Booking: Scheduling patient appointments and treatments
• Simulation: Both conventional and CT-based virtual simulation
• Document Import; reporting
• Image Management: Import/export of all image data associated with on-treatment verification; CTs, CBCTs, 2-D planar kV and MV images, CR, simulator; image processing and registration (set-up error) analysis; random and systematic set-up error analysis and trending; margin calculations; NAL protocols
• Radiotherapy (radiation oncology) process check lists (e.g. Mosaiq QCLs and IQ Scripts)

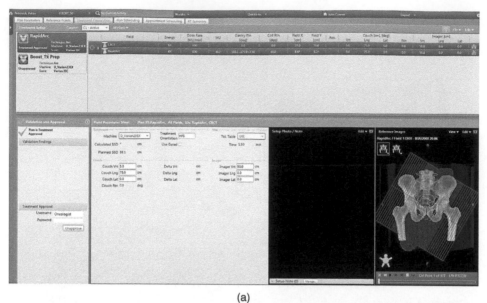

(a)

(b)

FIGURE 13.2 Typical screenshots of the OMS (OIS) software from (A) Varian's ARIA and (B) Elekta's MOSAIQ. (A) shows the treatment preparation page with the reference image (DRR) for a treatment field, functionality that is available across the network. (B) shows the integration of full 3-D computerized treatment planning into the OMS environment, so its availability is also across the network, including the linac control area.

Image (A) (top) courtesy of Varian Medical Systems; image (B) (bottom) courtesy of Elekta.

As the functionality has grown, the analytical aspects in particular of on-treatment imaging have migrated from the stand-alone EPID workstations described earlier to a fully centralised data and image database as part of the OMS with the benefits of remote and distributed image analysis and verification for off-line work. Images and information can viewed and interacted with from anywhere within a hospital network, and even across hospitals (e.g. in the UK for integrating satellite centres with the main cancer centre site; a means for increasing radiotherapy capacity) (Kirby et al. 2006a; Yu et al. 2010; Davies et al. 2011; IAEA 2013a; RCR 2013; Shakeshaft et al. 2014; BIR 2018). Algorithms for off-line protocols from individual patient data can now be applied across linacs for specific clinical sites in order to assess particular techniques (e.g. immobilisation and set-up) and derive target volume margins more easily from population data across a whole department (including Satellite centres). Some manufacturers build this capability into the OMS software; some hospitals have written bespoke software to interrogate, data-mine, and analyse set-up error data from the central OMS database or a separate imaging database.

With the movement of this on-treatment imaging functionality into the OMS, the onus on commissioning the hardware and software on a single new machine now extends onto the OMS as well; this could be especially frequent when, as a propriety-based system on common (e.g. Windows) architecture, software upgrades and patches are required throughout the year. As mentioned in section 13.2, commissioning is needed before first clinical use in a brand new networked configuration; but one must also consider the retesting required following software upgrades and patches that could potentially change functionality either deliberately (through better designs/algorithms) and/or inadvertently (through bugs and unseen conditional responses between functional coded units).

To this extent, the OMS should also be commissioned/tested for its on-treatment imaging and geometric verification functionality and capability. These tests could include, for example:

- Testing of algorithms such as are needed for functionality (image registration and analysis) performed on both the local on-treatment imaging acquisition system and on the software within the OMS, reconciling any differences that there might be (particularly if the two are from different manufacturers).

- Tests performed for image scaling, aspect ratio, orientation, and image registration results, as introduced in the section 13.2, could be executed on images within the OMS. These could be performed with either digital images in a library, or with real images of a phantom, or with a combination of images.

- End-to-end tests with a phantom (e.g. anthropomorphic) examining the full patient pathway from CT scanning => TPS => OMS => linac => OMS.

- Secure passage of set-up correction results and couch movements from linac on-treatment acquisition system to OMS, especially if the OMS will be used to analyse systematic and random set-up errors for the purpose for margin calculations.

In most cases, the initiation and acquisition of on-treatment imaging is still through local hardware on the linac, with hardwire links to, for example, remote couch movements for on-line work. But the passage of imaging data between local acquisition system and the database in, say, the OMS must be examined for the integrity of data transfer between the two. The origin of reference data (e.g. DRRs for 2-D imaging and 3-D and 4-D CT scans for 3-D/4-D on-treatment imaging) must also be considered for both integrated and third-party treatment planning systems and software.

References

Aaltonen, P., Brahme, A., Lax, I., et al. 1997. Specification of dose delivery in radiation therapy. Recommendations by the Nordic Association of Clinical Physics (NACP). *Acta oncologica.* 36(s10):1–32.

Abacioglu, U., Ozen, Z., Yilmaz, M., et al. 2014. Critical appraisal of RapidArc radiosurgery with flattening filter free photon beams for benign brain lesions in comparison to GammaKnife: A treatment planning study. *Radiation oncology.* 9:119.

Accuray website. *Accuray Inc: Innovating patient-first cancer treatment.* Available at: https://www.accuray.com/ (accessed on 3rd December 2018).

Acharya, S., Fischer-Valuck, B. W., Kashani, R. 2016a. Online magnetic resonance image guided adaptive radiation therapy: First clinical applications. *International journal of radiation oncology, biology, physics.* 94(2):394–403.

Acharya, S., Fischer-Valuck, B. W., Mazur, T. R. 2016b. Magnetic resonance image guided radiation therapy for external beam accelerated partial-breast irradiation: Evaluation of delivered dose and intrafractional cavity motion. *International journal of radiation oncology, biology, physics.* 96(4):785–792.

Ackerly, T., Lancaster, C. M., Geso, M. 2011. Clinical accuracy of ExacTrac intracranial frameless stereotactic system. *Medical physics.* 38(9):5040–5048.

AdvancedOncotherapy website. *First LIGHT system: In central London.* Available at: https://www.avoplc.com/Our-Technology/Toward-the-First-Installation-of-LIGHT (accessed on 3rd December 2018).

Aerts, H., Lambin, P., and De Ruysscher, D. 2010. FDG for dose painting: A rational choice. *Radiotherapy & oncology.* 97(2):163–164.

Ahmad, R., Bondar, L., Voet, P., et al. 2013. A margin-of-the-day online adaptive intensity-modulated radiotherapy strategy for cervical cancer provides superior treatment accuracy compared to clinically recommended margins: A dosimetric evaluation. *Acta oncologica.* 52(7):1430–1436.

Ahmed, S. K., Broan, P. D., Foote, R. L. 2018. Protons vs photons for brain and skull base tumors. *Seminars in radiation oncology.* 28(2):97–107.

Ahunbay, E. E., Ates, O., Li, X. A. 2016. An online replanning method using warm start optimization and aperture morphing for flattening-filter-free beams. *Medical physics.* 43(8):4575–4584.

Aird, E. G. A. 2004. Second cancer risk, concomitant exposures and IRMER(2000). *British journal of radiology.* 77(924):983–985.

Aird, E. G. A., Conway, J. 2002. CT simulation for radiotherapy treatment planning. *British journal of radiology.* 75:937–949.

Akimoto, M., Nakamura, M., Miyabe, Y., et al. 2015. Long-term stability assessment of a 4D tumor tracking system integrated into a gimbaled linear accelerator. *Journal of applied clinical medical physics.* 16(5):373–380.

Alander, E., Visapaa, H., Kouri, M., et al. 2014. Gold seed fiducials in analysis of linear and rotational displacement of the prostate bed. *Radiotherapy & oncology*. 110:256–260.

Alexander III, E., Moriarty, T. M., Davis, R. B. Stereotactic radiosurgery for the definitive, noninvasive treatment of brain metastases. *Journal of the national cancer institute*. 1995, 87(1):34–40

Allen, A., Pawlicki, T., Dong, L., et al. 2012. An evidence based review of proton beam therapy: The report of ASTRO's emerging technology committee *Radiotherapy & oncology*. 103:8–11.

American Association of Physicists in Medicine (AAPM website). Available at: https://w3.aapm.org/my_aapm/index.php (accessed on 3rd December 2018).

American Society for Radiation Oncology (ASTRO) website. Available at: https://www.astro.org (accessed on 3rd December 2018).

Anamalayil, S., Brady, L., Grover, S., et al. 2017. Treatment of cervical cancer with a prototype flattening filter-free straight through linac with fast jawless MLC collimator. *International journal of radiation oncology, biology, physics*. 99(2):E633.

Anantham, D., Feller-Kopman, D., Shanmugham, L. N., et al. 2007. Electromagnetic navigation bronchoscopy guided fiducial placement for robotic stereotactic radiosurgery of lung tumors—a feasibility study. *Chest*. 132(3):930–935.

Antonuk, L. E. 2002. Electronic portal imaging devices: A review and historical perspective of contemporary technologies and research. *Physics in medicine and biology*. 47:R31–R65.

Antypas, C., Pantelis, E. 2008. Performance evaluation of a CyberKnife G4 image-guided robotic stereotactic radiosurgery system. *Physics in medicine and biology*. 53:4697–4718.

Apicella, G., Loi, G., Torrente, S., et al. 2016. Three-dimensional surface imaging for detection of intra-fraction setup variations during radiotherapy of pelvic tumors. *La radiologia medica*. 121(10):805–810.

Armstrong, J. G. 1998. Target volume definition for three-dimensional conformal radiation therapy of lung cancer. *The British journal of radiology*. 71:587–594.

ARO website. ASTRO '18 RefleXion unveils biology-guided radiotherapy system. Oct. 20, 2018. https://appliedradiationoncology.com/articles/astro-18-reflexion-unveils-biology-guided-radiotherapy-system (accessed on 3rd December 2018).

Arumugam, S., Jameson, M. G., Xing, A. et al. 2013. An accuracy assessment of different rigid body image registration methods and robotic couch positional corrections using a novel phantom. *Medical physics*. 40(3):031701-1–9.

ASN, Autorite De Surete Nucleaire. *Management guidelines for safety and quality in radiotherapy: User guide for the implementation of quality assurance in radiotherapy*. Paris, France: ASN, 2009. (www.asn.fr)

Australasian College of Physical Scientists and Engineers in Medicine (ACPSEM) et al. 2013. *Radiation Oncology Practice Standards: New Zealand*. Available at: https://www.acpsem.org.au/documents/item/71

Aznar, M. C., Warren, S., Hoogeman, M., et al. 2018. The impact of technology on the changing practice of lung SBRT. *Physica medica*. 47:129–138.

Balog, J., Soisson, E., 2008. Helical tomotherapy quality assurance. *International journal of radiation oncology, biology, physics*. 71(1):S113–S117.

Balter, J., Wright, J., Newell, L., et al. 2005 .Accuracy of a wireless localization system for radiotherapy. *International journal of radiation oncology biology physics*. 61(3):933–937.

Balter, P., Netherton, T., Li, Y., et al. 2017. Dose considerations of IGRT sing MV projection and MV CBCT on a prototype linear accelerator. *Radiotherapy & oncology*. 123 (Suppl 1):S510–S511.

Barney, B. M., Lee, R. J., Handrahan, D., et al. 2011. Image-guided radiotherapy (IGRT) for prostate cancer comparing kV imaging of fiducial markers with cone beam computed tomography (CBCT). *International journal of radiation oncology, biology, physics*. 80(1):301–305.

Baroni, G., Ferrigno, G., Pedotti, A. 1998. Implementation and application of real-time motion analysis based on passive markers. *Medical & biological engineering & computing.* 36(6):693–703.

Bartling, S., Sharp, G. C., Pauwels, R., et al. *ICRP Report No. 129: Radiological protection in cone beam computed tomography (CBCT).* Ottawa, Canada: ICRP, 2015.

Bates, A. M., Scaife, J., Tudor, G. S. J., et al. 2013. Image guidance protocols: Balancing imaging parameters against scan time. *British journal of radiology.* 86:20130385.

Batin, E., Depauw, N., MacDonald, S., et al. 2016. Can surface imaging improve the patient setup for proton postmastectomy chest wall irradiation? *Practical radiation oncology.* 6(6):e235–e241.

Bauman, G., Yartsev, S., Rodrigues, G., et al. 2007. A prospective evaluation of helical tomotherapy. *International journal of radiation oncology, biology, physics.* 68(2):632–641.

Beavis, A. W. 2004. Is tomotherapy the future of IMRT? *British journal of radiology.* 77:285–295.

Becker, M., Zaidi, H. 2014. Imaging in head and neck squamous cell carcinoma: The potential role of PET/MRI. *British journal of radiology.* 87:20130677.

Bedford, J., Hanson, I., Hansen, V. 2018. Comparison of forward- and back-projection in vivo EPID dosimetry for VMAT treatment of the prostate. *Physics in medicine and biology.* 63:025008 (16 pp).

Bel, A., Van Herk, M., Bartelink, H., et al. 1993. A verification procedure to improve patient set-up accuracy using portal images. *Radiotherapy & oncology.* 29(2):253–260.

Berkovic, P., Paelinck, L., Gulyban, A., et al. 2017. Adaptive radiotherapy for locally advanced non-small cell lung cancer: Dosimetric gain and treatment outcome prediction. *Acta oncologica.* 56(11):1656–1659.

Berrang, T. S., Truong, P. T., Popescu, C. 2009. 3-D ultrasound can contribute to CT to define the target for partial breast radiotherapy. *International journal of radiation oncology, biology, physics.* 73(2):375–383.

Bert, C., Metheany, K., Dopple, K., et al. 2005. A phantom evaluation of a stereo-vision surface imaging system for radiotherapy patient setup. *Medical physics.* 32(9):2753–2762.

Bert, C., Metheany, K., Doppke, K., et al. 2006. Clinical experience with a 3-D surface patient setup system for alignment of partial-breast irradiation patients. *International journal of radiation oncology biology physics.* 64(4):1265–1274.

Biggs, P., Goitein, M., Russell, M. 1985. A diagnostic x-ray field verification device for a 10 MV linear accelerator. *International journal of radiation oncology biology physics.* 11:635–643.

Bijderkerke, P., Verellen, D., Tournel, K., et al. 2008. Tomotherapy: Implications on daily workload and scheduling patients. *Radiotherapy & oncology.* 86:224–230.

BIR. 2014. *Meeting the current and future workforce challenges for patient care in a changing context.* Available at https://www.bir.org.uk/media/231974/19_may_programme_for_website.pdf (accessed on 3rd December 2018).

BIR. 2016. *VERT: An (Online) educational study day for present and new users.* Available at https://www.bir.org.uk/education-and-events/vert-on-demand-content/ (accessed on 3rd December 2018).

BIR. 2018. *Success of satellite radiotherapy and remote working in radiotherapy.* Available at https://www.bir.org.uk/media/372359/success_of_satellites_programme.pdf (accessed on 3rd December 2018).

Bissonette, J. P., Balter, P. A., Dong, L., et al. 2012. Quality assurance for image guided radiation therapy utilizing CT-based technologies: A report of the AAPM TG-179. *Medical physics.* 39(4):1946–1963.

Blanchard, P., Gun, G., Lin, A., et al. 2018. Proton therapy for head and neck cancers. *Seminars in radiation oncology.* 28(1):53–63.

Boda-Heggemann, J., Kohler, F., Kupper, B., et al. 2008. Accuracy of ultrasound-based (BAT) prostate-repositioning: A three-dimensional on-line fiducial-based assessment with cone-beam computed tomography. *International journal of radiation oncology biology physics.* 70(4):1247–1255.

Bogdanich, W. 2010. The radiation boom—radiation offers new cures, and ways to do harm. *New York Times.* 23rd January 2010. Available at: http://www.nytimes.com/2010/01/24/health/24radiation.html

Bohoudi, O., Bruynzeel, A. M. E., Senan, S. 2017. Fast and robust online adaptive planning in stereotactic MR-guided adaptive radiation therapy (SMART) for pancreatic cancer. *Radiotherapy & oncology.* 125(3):439–444.

Bol, G., Lagendijk, J., Raaymakers, B. 2013. Virtual couch shift (VCS): Accounting for patient translation and rotation by online IMRT re-optimization. *Physics in medicine and biology.* 58(9):2989.

Bortfeld, T., Chen, G. 2004. Intrafractional organ motion and its management. *Seminars in radiation oncology.* 14(1):1.

Bortfeld, T., Van Herk, M., Jiang, S. B. 2002. When should systematic patient positioning errors in radiotherapy be corrected? *Physics in medicine and biology.* 47:N297.

Bourland, D. (Ed.) 2012. *Image guided radiation therapy.* Boca Raton, FL: CRC Press, Taylor and Francis Group.

Bova, F., Buatti, J., Friedman W., et al. 1997. The university of Florida frameless high-precision stereotactic radiotherapy system. *International journal of radiation oncology biology physics.* 38(4):875–882.

Boyer, A., Antonuk, L., Fenster, A., et al. 1992. A review of electronic portal imaging devices (EPIDs). *Medical physics.* 19(1):1–16.

Bradford, S., Kirby, M. C. 2017. *A service evaluation of the use of VMAT with CBCT imaging for Ca Cervix patients—evaluating the margin for target volume coverage with bone-registration matching.* Poster (e119) presented at UKRCO 2017, UK Radiation Oncology Congress, 12–14 June 2017, Manchester, UK.

Bradley, J. Bae, K. Choi, N. et al. 2012. A phase II comparative study of gross tumor volume definition with or without PET/CT fusion in dosimetric planning for non–small-cell lung cancer (nsclc): Primary analysis of radiation therapy oncology group (RTOG) 0515. *International journal of radiation oncology, biology, physics.* 82(1):435–441.

Brady, L., Scheuermann, R., Anamalayil, S., et al. 2017. Robustness of extended field cervical target optimization techniques to isocenter offsets with a prototype fast jawless MLC system. *International journal of radiation oncology, biology, physics.* 99(2):E642.

Bragstad, S., Flatebo, M., Natvig, G. K., et al. 2018. Predictors of quality of life and survival following Gamma Knife surgery for lung cancer brain metastases: A prospective study. *Journal of neurosurgery.* 129:71–83.

Brahme, A., Nyman, P., Skatt, B. 2008. 4-D laser camera for accurate patient positioning, collision avoidance, image fusion and adaptive approaches during diagnostic and therapeutic procedures. *Medical physics.* 35(5):1670–1681.

Braide, K., Lindencrona, U., Welinder, K., et al. 2018. Clinical feasibility and positional stability of an implanted wired transmitter in a novel electromagnetic positioning system for prostate cancer radiotherapy. *Radiotherapy & oncology.* 128(2):336–342.

Bridge, P., Giles, E., Williams, A. et al. 2017. International audit of virtual environment for radiotherapy training usage. *Journal of radiotherapy in practice.* 16:375–382.

British Institute of Radiology (BIR) website. Available at: https://www.bir.org.uk/(accessed on 3rd December 2018).

Brock, K., Dawson, L. 2014. Point: Principles of magnetic resonance imaging integration in a computed tomography–based radiotherapy workflow. *Seminars in radiation oncology.* 24:169–174.

Brock, K. K., Mutic, S., McNutt, T., et al. 2017. Use of image registration and fusion algorithms and techniques in radiotherapy: Report of the AAPM Radiation Therapy Committee Task Group No. 132. *Medical physics.* 44(7):e43–e75.

Brouwer, C. L., Steenbakkers, R. J. H. M., Langendijk, J. A., et al. 2015. Identifying patients who may benefit from adaptive radiotherapy: Does the literature on anatomic and dosimetric changes in head and neck organs at risk during radiotherapy provide information to help? *Radiotherapy & oncology.* 115:285–294.

Buchbender, C., Heusner, T., Laurenstein, T., et al. 2012. Oncologic PET/MRI, Part 1: Tumors of the brain, head and neck, chest, abdomen, and pelvis. *Journal of nuclear medicine.* 53:928–938.

Buitenhuis, H., Diblen, F., Brzezinski, K., et al. 2017. Beam-on imaging of short-lived positron emitters during proton therapy. *Physics in medicine and biology.* 62:4654–4672.

Burghelea, M., Verellen, D., Dhont, J., et al. 2017. Treating patients with Dynamic Wave Arc: First clinical experience. *Radiotherapy & oncology.* 122:347–351.

Burghelea, M., Verellen, D., Poels, K., et al. 2016. Initial characterization, dosimetric benchmark and performance validation of Dynamic Wave Arc. *Radiation oncology.* 11:63.

Burmeister, J., Chen, Z., Chetty, I. J., et al. 2016. The American Society for Radiation Oncology's 2015 core physics curriculum for radiation oncology residents. *International journal of radiation oncology, biology, physics.* 95(4):1298–1303.

Burnet, N. G., Adams, E., Fairfoul, J., et al. 2010. Practical aspects of implementation of helical tomotherapy for intensity-modulated and image-guided radiotherapy. *Clinical oncology (Royal College of Radiologists).* 22(4):294–312.

Byrne, J. 2017. Guidance on MGTI: Implementation in radiotherapy. *SCOPE.* June:34–36. York, UK: IPEM.

Cadieux, C. L., DesRosiers, C., McMullen, K. 2016. Risks of secondary malignancies with heterotopic bone radiation therapy for patients younger than 40 years. *Medical dosimetry.* 41:212–215.

Callahan, J., Kron, T., Schneider-Kolsky, M. 2013. Validation of a 4D-PET maximum intensity projection for delineation of an internal target volume. *International journal of radiation oncology, biology, physics.* 86:749–754.

Caporaso, G. J., Mackie, T., Sampayan, S., et al. 2008. A compact linac for intensity modulated proton therapy based on a dielectric wall accelerator. *Physica medica.* 24(2):98–101.

Care Quality Commission. 2018. *CQC's enforcement policy for the Ionising Radiation (Medical Exposure) Regulations 2017.* London, UK: Care Quality Commission.

Castadot, P., Lee, J., Geets, X., et al. 2010. Adaptive radiotherapy of head and neck cancer. *Seminars in radiation oncology.* 20:84–93.

Cervino, L., Detorie, N., Taylor, M., et al. 2012. Initial clinical experience with a frameless and maskless stereotactic radiosurgery treatment. *Practical radiation oncology.* 2:54–62.

Cevik, M., Shirvani Ghomi, P., Aleman, D., et al. 2018. Modeling and comparison of alternative approaches for sector duration optimization in a dedicated radiosurgery system. *Physics in medicine and biology.* 63:155009 (16 pp.).

Chan, M., Grehn, M., Cremers, F., et al. 2017. Dosimetric implications of residual tracking errors during robotic SBRT of liver metastases. *International journal of radiation oncology, biology, physics.* 97(4):839–848.

Chan, M. F., Yang, J., Song, Y., et al. 2011. Evaluation of imaging performance of major image guidance systems. *Biomedical imaging and intervention journal.* 7(2):e11.

Chang, D. W., Cheetham, L., te Marvelde, L., et al. 2014. Risk factors for radiotherapy incidents and impact of an online electronic reporting system. *Radiotherapy & oncology.* 112:199–204.

Chatterjee, S., Mott, J. H., Smyth J., et al. 2011. Clinical challenges in the implementation of a tomotherapy service for head and neck cancer patients in a regional UK radiotherapy centre. *British journal of radiology.* 84:358–366.

Chaurasia, A. R., Sun, K. J., Premo, C., et al. 2018. Evaluating the potential benefit of reduced planning target volume margins for low and intermediate risk patients with prostate cancer using real-time electromagnetic tracking. *Advances in radiation oncology.* 3:630–638.

Chen, A., Cao, M., Hus, S., et al. 2017. Magnetic resonance imaging guided reirradiation of recurrent and second primary head and neck cancer. *Advances in radiation oncology.* 2(2):167–175.

Chen, A., Daly, M., Cui, J., et al. 2014. Clinical outcomes among patients with head and neck cancer treated by intensity-modulated radiotherapy with and without adaptive replanning. *Head and neck.* 36:1541–1546.

Chen, A. M., Farwell, D. G., Luu, Q., et al. 2011. Evaluation of the planning target volume in the treatment of head and neck cancer with intensity-modulated radiotherapy: What is the appropriate expansion margin in the setting of daily image guidance? *International journal of radiation oncology, biology, physics.* 81(4):943–949.

Chen, G., Riboldi, M., Gierga, D., et al. 2005. Clinical implementation of IGRT techniques. *Radiotherapy & oncology.* 76(S2):S10.

Chen, J., Morin, O., Aubin, M., et al. 2006. Dose-guided radiation therapy with megavoltage cone-beam CT. *British journal of radiology.* 79(1):S87–S98.

Chen, M., Chao, E., Lu, W., 2013. Quantitative characterization of tomotherapy MVCT dosimetry. *Medical dosimetry.* 38:280–286.

Chetty, I., Paysan, P., Siddiqui, F., et al. 2017. Improving CBCT image quality for daily image guidance of patients with head/neck and prostate cancer. *Radiotherapy & oncology.* 123(Suppl 1):S491–S492.

Cho, J. H., Lee, C-G., Kang, D. R. 2009. Positional reproducibility and effects of a rectal balloon in prostate cancer radiotherapy. *Journal of Korean medical science.* 24:894–903.

Choi, C. H., Park, S-Y., Kim, J-I., et al. 2017. Quality of tri-Co-60 MR-IGRT treatment plans in comparison with VMAT treatment plans for spine SABR. *British journal of radiology.* 90:20160652.

Choudhury, A., Budgell, G., Mackay, R., et al. 2017. The future of image-guided radiotherapy. *Clinical oncology (Royal College of Radiologists).* 29:662–666.

Chung, M. J., Kim, S. H., Lee, J. H., et al. 2015. A dosimetric comparative analysis of TomoDirect and three-dimensional conformal radiotherapy in early breast cancer. *Journal of breast cancer.* 18(1):57–62.

Chung, P. W., Haycocks, T., Brown, T., et al. 2004. On-line aSi portal imaging of implanted fiducial markers for the reduction of interfraction error during conformal radiotherapy of prostate carcinoma. *International journal of radiation oncology, biology, physics.* 60(1):329–334.

Clark, B. G., Brown, R. J., Ploquin, J. L., et al. 2010. The management of radiation treatment error through incident learning. *Radiotherapy & oncology.* 95:344–349.

Clemente, S., Chiumento, C., Fiorentino, A., et al. 2013. Is ExacTrac x-ray system an alternative to CBCT for positioning patients with head and neck cancers? *Medical physics.* 40(11):111725-1-111725-6

Coffey, M., Cuningham, Y., Holmberg, O. et al. *Accident reporting system: The ROSIS experience.* International Conference on Modern Radiotherapy: Advances and Challenges in Radiation Protection of Patients. Versailles, France, Dec 2–4, 2009. Available at: https://inis.iaea.org/collection/NCLCollectionStore/_Public/43/043/43043546.pdf?r=1&r=1 (accessed on 3rd December 2018).

Court, L., McCarroll, R., Kisling, K., et al. 2017. Full automation of radiation therapy treatment planning. *Radiotherapy & oncology.* 123(Suppl 1):S439–S440.

Cozzi, L., Fogliata, A., Thompson, S., et al. 2018. Critical appraisal of the treatment planning performance of volumetric modulated arc therapy by means of a dual layer stacked multileaf collimator for head and neck, breast, and prostate VMAT HnN, breast prostate. *Technology in cancer research & treatment.* 17:1–11.

CQC, Care Quality Commission web page. Available at: https://www.cqc.org.uk/ (accessed on 3rd December 2018).

Crook, J. M., Raymond, Y., Salhani, D., et al. 1995. Prostate motion during standard radiotherapy as assessed by fiducial markers. *Radiotherapy & oncology.* 37:35–42.

Cunningham, J., Coffey, M., Knoos, T., Holmberg, O. 2010. Radiation oncology safety information system (ROSIS)—profiles of participants and the first 1074 incident reports. *Radiotherapy & oncology.* 97:601–607.

Cvetkova, J., O'Donovan, T., Craig, A., et al. 2010. Radiation therapists' compliance to a palliative imaging protocol: A case report. *Journal of radiotherapy in practice*. 17(2):253–256.

D'Ambrosio, D. J., Bayouth, J., Chetty, I. J., et al. 2012. Continuous localization technologies for radiotherapy delivery: Report of the American Society for Radiation Oncology Emerging Technology Committee. *Practical radiation oncology*. 2:145–150.

Das, S., Liu, T., Jani, A. B., et al. 2014. Comparison of image-guided radiotherapy technologies for prostate cancer. *American journal of clinical oncology*. 37(6):616–623.

Datta, N., David, R., Gupta, R., et al. 2008. Implications of contrast-enhanced CT-based and MRI-based target volume delineations in radiotherapy treatment planning for brain tumors. *Journal of cancer research and therapeutics*. 4(1):9–13.

Davies, J., Oliver, K., Chatterton, S., et al. 2011. Delivering radiotherapy in a fully networked paperless satellite department. *Clinical oncology. (Royal College of Radiologists)* 23(3):S35.

De Blanck, S., Rydhog, J., Larsen, K., et al. 2018. Long term safety and visibility of a novel liquid fiducial marker for use in image guided radiotherapy of non-small cell lung cancer. *Clinical and translational radiation oncology*. 13:24–28.

de Boer, H. C., Heijmen, B. J. 2007. eNAL: An extension of the NAL setup correction protocol for effective use of weekly follow-up measurements. *International journal of radiation oncology, biology, physics*. 67(5):1586–1595.

de Boer, H. C., Van Sornsen de Koste, J. R., Creutzberg, C. L., et al. 2001. Electronic portal image assisted reduction of systematic set-up errors in head and neck irradiation. *Radiotherapy & oncology*. 61(3):299–308.

de Boer, H. C. J., Heijmen, B. J. M. 2001. A protocol for the reduction of systematic patient setup errors with minimal portal imaging workload. *International journal of radiation oncology, biology, physics*. 50(5):1350–1365.

de Boer, H. C. J., Van Os, M. J. H., Jansen, P. P., et al. 2005. Application of the No Action Level (NAL) protocol to correct for prostate motion based on electronic portal imaging of implanted markers. *International journal of radiation oncology, biology, physics*. 61(4):969–983.

De Los Santos, J., Popple, R., Agazaryan, N., et al. 2013. Image guided radiation therapy (IGRT) technologies for radiation therapy localization and delivery. *International journal of radiation oncology, biology, physics*. 87(1):33–45.

De Rooij, M., Hamoen, E., Futterer, J., et al. 2014. Accuracy of multiparametric MRI for prostate cancer detection: A meta-analysis. *American journal of roentgenology*. 202(2):343–351.

De Ruysscher, D., Chang, J. Y., 2013. Clinical controversies: Proton therapy for thoracic tumors. *Seminars in radiation oncology*. 23:115–119.

De Ruysscher, D., Lodge, M., Jones, B., et al. 2012. Charged particles in radiotherapy: A 5-year update of a systematic review *Radiotherapy & oncology*. 103:5–7.

De Ruysscher, D., Wanders, S., Minken, A., et al. 2005. Effects of radiotherapy planning with a dedicated combined PET-CT-simulator of patients with non-small cell lung cancer on dose limiting normal tissues and radiation dose-escalation: A planning study. *Radiotherapy & oncology*. 77:5–10.

Dean, J. C., Tudor, G. S. J., Mott, J., et al. 2013. Multi-centre experience of implementing image-guided intensity-modulated radiotherapy using the TomoTherapy platform. *Radiography*. 19:270–273.

Deegan, T., Owen, R. Holt, T., et al. 2015. Assessment of cone beam CT registration for prostate radiation therapy: Fiducial marker and soft tissue methods. *Journal of medical imaging and radiation oncology*. 59:91–98.

Degiovanni, A. 2014. *Accelerators for hadron therapy*. Spain: IVICFA Valencia. Available at: https://indico.ific.uv.es/event/2344/contributions/2263/attachments/1905/2146/Accelerators-Hadron-Therapy_2014_Degiovanni.pdf (accessed on 3rd December 2018).

Degiovanni, A., Amaldi, U. 2015. History of hadron therapy accelerators. *Physica medica*. 31:322–332.

Degiovanni, A., Stabile, P., Ungaro, D. 2016. *LIGHT: A linear accelerator for proton therapy.* Proceedings of NAPAC2016, Chicago, Illinois USA, 1282–1286. Available at http://accelconf. web.cern.ch/AccelConf/napac2016/papers/frb1io02.pdf (accessed on 3rd December 2018).

Delso, G., Furst, S., Jakoby, B., et al. 2011. Performance measurements of the Siemens mMR integrated whole-body PET/MR scanner. *Journal of nuclear medicine.* 52:1914–1922.

Dendooven, P., Buitenhuis, H., Diblen, F., et al. 2015. Short-lived positron emitters in beam-on PET imaging during proton therapy. *Physics in medicine and biology.* 60:8923–8947.

Department of Health (DoH). 2017. *Guidance on investigation and notification of medical exposures much greater than intended.* Published 16 January 2017. London, UK.

Depauw, N., Dias, M., Rosenfeld, A., et al. 2014. Ion radiography as a tool for patient set-up & image guided particle therapy: A Monte Carlo study. *Technology in cancer research & treatment.* 13(1):69–76.

Depuydt, T., Verellen, D., Haas, O., et al. 2011. Geometric accuracy of a novel gimbals based radiation therapy tumour tracking system. *Radiotherapy & oncology.* 98:365–372.

Dieterich, S., Cavedon, C., Chuang, C. F., et al. 2011. Report of AAPM TG 135: Quality assurance for robotic radiosurgery. *Medical physics.* 38(6):2914–2936.

Dieterich, S., Ford, E., Pavord, D., et al. 2016a. Image guidance and localization technologies for radiotherapy. In: *Practical Radiation Oncology Physics.* Amsterdam, Netherlands: Elsevier.

Dieterich, S., Ford, E., Pavord, D., et al. 2016b. Clinical aspects of image guidance and localization in radiotherapy. In: *Practical Radiation Oncology Physics.* Amsterdam, Netherlands: Elsevier.

Ding, G. 2015. Characteristics of x-rays from ExacTrac and patient dose from imaging procedures. *Medical physics.* 42:3268.

Ding, G. X., Alael, P., Curran, B., et al. 2018. Image guidance doses delivered during radiotherapy: Quantification, management, and reduction: Report of the AAPM Therapy Physics Committee Task Group 180. *Medical physics.* 45(5):e84–e99.

Ding, G. X., Coffey, C. W. 2008. Is it time to include imaging guidance doses in the reportable total radiation doses of radiotherapy patients? *International journal of radiation oncology, biology, physics.* 72:S145–S146.

Ding, G. X., Coffey, C. W. 2009. Radiation dose from kilovoltage cone beam computed tomography in an image-guided radiotherapy procedure. *International journal of radiation oncology, biology, physics.* 73:610–617.

Ding, G. X., Duggan, D. M., Coffey, C. 2007. Additional doses to critical organs from CBCT scan in image guided radiation therapy. *International journal of radiation oncology, biology, physics.* 69:S45.

Ding, G. X., Duggan, D. M., Coffey, C. W. 2008. Accurate patient dosimetry of kilovoltage cone-beam CT in radiation therapy. *Medical physics.* 35:1135–1144.

Ding, G. X., Duggan, D. M., Coffey, C. W., et al. 2007. A study on adaptive IMRT treatment planning using kV cone-beam CT. *Radiotherapy & oncology.* 85:116–125.

Ding, G. X., Munro, P. 2013. Radiation exposure to patients from image guidance procedures and techniques to reduce the imaging dose. *Radiotherapy & oncology.* 108:91–98.

DNV GL. 2015. *DNV-GL Guidance Document—ISO 9001:2015: Quality management systems— requirements.* Hovik, Norway: DNV GL AS. Available at https://www.dnvgl.co.uk/Images/ISO9001_DNVGL_2015_tcm12-31224.pdf (accessed on 3rd December 2018).

Dobler, B., Mai, S., Wolff, D., et al. 2006. Evaluation of possible prostate displacement induced by pressure applied during transabdominal ultrasound image acquisition. *Strahlentherapie und onkologie.* 182(4):240–246.

DoH, UK Department of Health. 1990. *Quality Systems for Medical Equipment 1990: Good Manufacturing Practice.* London, UK: The Stationery Office.

Dong, L., Boyer, A. L. 1995. An image correlation procedure for digitally reconstructed radiographs and electronic portal images. *International journal of radiation oncology, biology, physics.* 33(5):1053–1060.

Drzezga, A., Souvatzoglou, M., Fiber, M., et al. 2012. First clinical experience with integrated whole-body PET/MR: Comparison to PET/CT in patients with oncologic diagnoses. *Journal of nuclear medicine.* 53:845–855.

Duffton, A., Harrow, S., Lamb, C., et al. 2016. An assessment of cone beam CT in the adaptive radiotherapy planning process for non-small-cell lung cancer patients. *British journal of radiology.* 89:20150492.

Dufton, A., McNee, S., Muirhead, R., et al. 2012. Clinical commissioning of online seed matching protocol for prostate radiotherapy. *The British journal of radiology.* 85:e1273–e1281.

Durante, M., Debus, J., 2018. Heavy charged particles: Does improved precision and higher biological effectiveness translate to better outcome in patients? *Seminars in radiation oncology.* 28(2):160–167.

Eaton, D. J., Byrne, J. P., Cosgrove, V. P., et al. 2018. Unintended doses in radiotherapy—over, under and outside? *British journal of radiology.* 91:20170863.

Elekta. 2011. *Clarity, soft tissue visualization.* Clarity product brochure available at https://5.imimg. com/data5/UF/DB/MY-3883832/elekta-clarity-treatment-solutions.pdf (accessed on 3rd December 2018).

Elekta. 2015. *Design and performance characteristics of a Cone Beam CT system for Leksell Gamma Knife® Icon™.* Available at: https://www.careforthebrain.com/

Elekta. 2018a. *MOSAIQ radiation oncology.* Product brochure. Available at: https://www.elekta. com/software-solutions/care-management/mosaiq-radiation-oncology/ (accessed on 3rd December 2018).

Elekta. 2018b. *MOSAIQ Radiation Oncology Information System: Confidence for you and your patients.* Stockholm, Sweden: Elekta AB.

Elekta. 2018c. *Elekta AB: Leksell Gamma Knife® Perfexion.* Available at: https://www.elekta.com/ radiosurgery/leksell-gamma-knife-perfexion/ (accessed on 3rd December 2018).

Emmerson, J., Clark, J., Mansell, C. 2012. Quality management in radiotherapy. In: Symonds, P., Deehan, C., Mills, J., et al. (eds.), *Walter and Miller's Textbook of Radiotherapy: Radiation Physics, Therapy and Oncology.* London, UK: Elsevier Churchill Livingstone, 2012.

Engelsman, M., Schwarz, M., Dong, L. 2013. Physics controversies in proton therapy. *Seminars in radiation oncology.* 23:88–96.

Erdi, Y. E., Rosenzweig, K., Erdi, A. K. 2002. Radiotherapy treatment planning for patients with non-small cell lung cancer using positron emission tomography (PET). *Radiotherapy & oncology.* 62:51–60.

Eriksen, J. G., Beavis, A. W., Coffey, M. A., et al. 2012. The updated ESTRO core curricula 2011 for clinicians, medical physicists and RTTs in radiotherapy/radiation oncology. *Radiotherapy & oncology.* 103:103–108.

European Society for Radiotherapy and Oncology (ESTRO) website. Available at: https://www. estro.org/ (accessed on 3rd December 2018).

European Union of Medical Specialists (EUMS). 2013. *Training requirements for the specialty of radiation oncology.* Vienna, Austria: ESTRO/UEMS.

Evans, P. M. 2008. Anatomical imaging for radiotherapy. *Physics in medicine and biology.* 53:R151–R191.

Ezzell, G., Burmeister, J., Doggan, N., et al. 2009. IMRT commissioning: Multiple institution planning and dosimetry comparisons, a report from AAPM task group 119. *Medical physics.* 36:5359–5373.

Fahimian, B., Soltys, S., Xing, L., et al. 2013. Evaluation of MLC-based robotic radiotherapy. *Medical physics.* 40:344.

Fallone, B. 2014. The rotating biplanar linac–magnetic resonance imaging system. *Seminars in radiation oncology.* 24:200–202.

Fallone, B., Murray, B., Rathee, S., et al. 2009. First MR images obtained during megavoltage photon irradiation from a prototype integrated linac-MR system. *Medical physics.* 36(6):2084–2088.

Fan, Q., Nanduri, A., Mazin, S., et al. 2012. Emission guided radiation therapy for lung and prostate cancers: A feasibility study on a digital patient. *Medical physics*. 39(11):7140–7152.

Fan, Q., Nanduri, A., Yang, J. et al. 2013. Toward a planning scheme for emission guided radiation therapy (EGRT): FDG based tumor tracking in a metastatic breast cancer patient. *Medical physics*. 40(8):081708-1–12.

Fan, Q., Zhu, L. 2010. Emission guided radiation therapy: A simulation study of treatment without margin. *Medical physics*. Su-HH-BRB-06: https://doi.org/10/1118/1.3469024

Fattori, G., Safai, S., Carmona, P., et al. 2017. Monitoring of breathing motion in image-guided PBS proton therapy: Comparative analysis of optical and electromagnetic technologies. *Radiation oncology*. 12:63.

Fenwick, J. D., Tome, W. A., Jaradat, H. A., et al. 2004. Quality assurance of a helical tomotherapy machine. *Physics in medicine and biology*. 49:2933–2953.

Fenwick, J. D., Tome, W. A., Soisson E. T. et al. 2006. Tomotherapy and other innovative IMRT delivery systems. *Seminars in radiation oncology*. 16:199–208.

Fernandes, A., Apisarnthanarax, S., Yin, L., et al. 2015. Comparative assessment of liver tumor motion using cine–magnetic resonance imaging versus 4-dimensional computed tomography. *International journal of radiation oncology, biology, physics*. 91(5):1034–1040.

Findlay, U., Best, H., Ottrey, M. 2016. Improving patient safety in radiotherapy through error reporting and analysis. *Radiography*. 22:S3–S11.

Fiorina, E., Ferrero, V., Pennazio, F., et al. 2018. Monte Carlo simulation tool for online treatment monitoring in hadrontherapy with in-beam PET: A patient study. *Physica medica*. 51:71–80.

Fiorino, C., Alongi, F., Broggi, S., et al. 2008. Physics aspects of prostate tomotherapy: Planning optimization and image-guidance issues. *Acta oncologica*. 47(7):1309–1316.

Fischer-Valuck, B., Henke, L., Green, O., et al. 2017. Two-and-a-half-year clinical experience with the world's first magnetic resonance image guided radiation therapy system. *Advances in radiation oncology*. 2:485–493.

Flanz, J., Bortfeld, T. 2013. Evolution of technology to optimize the delivery of proton therapy: The third generation. *Seminars in radiation oncology*. 23:142–148.

Fontanarosa, D., van der Meer, S., Bamber, J., et al. 2015. Review of ultrasound image guidance in external beam radiotherapy: I. Treatment planning and inter-fraction motion management. *Physics in medicine & biology*. 60:R77–R114.

Fontenot, J. D., Alkhatib, H., Garrett, J. A., et al. 2014. AAPM medical physics practice guideline 2.a: Commissioning and quality assurance of x-ray–based image-guided radiotherapy systems. *Journal of applied clinical medical physics*. 15(1):3–13.

Ford, E. C., Evans, S. B. 2018. Incident learning in radiation oncology: A review. *Medical physics*. 45(5):e100–e119.

Foroudi, F., Pham, D., Bressel, M., et al. 2014. Comparison of margins, integral dose and interfraction target coverage with image-guided radiotherapy compared with non-image-guided radiotherapy for bladder cancer. *Clinical oncology (Royal College of Radiologists)*. 26(8):497–505.

Foster, R. D., Pistenmaa, D. A., Solberg, T. D. 2012. A comparison of radiographic techniques and electromagnetic transponders for localization of the prostate. *Radiation oncology*. 7:101.

Francis, L., Sipahi, E. 2014. *Centralizing image and data management in radiation oncology with MOSAIQ Data Director™*. Stockholm, Sweden: Elekta AB.

Franco, P., Zeverino, M., Migliaccio, F., et al. 2014. Intensity-modulated and hypofractionated simultaneous integrated boost adjuvant breast radiation employing statics ports of tomotherapy (TomoDirect): A prospective phase II trial. *Journal of cancer research and clinical oncology*. 140:167–177.

Frank, S. J., Blanchard, P., Lee, J., et al. 2018. Comparing intensity-modulated proton therapy with intensity-modulated photon therapy for oropharyngeal cancer: The journey from clinical trial concept to activation. *Seminars in radiation oncology*. 28(2):108–113.

Fu, D., Kuduvalli, G. 2006. *Enhancing skeletal features in digitally reconstructed radiographs.* Proceedings of SPIE. 6144:846–851.

Fu, D., Kuduvalli, G. 2008. A fast, accurate, and automatic 2-D-3-D image registration for image-guided cranial radiosurgery. *Medical physics.* 35:2180–2194.

Fu, D., Kuduvalli, G., Maurer, C. R., et al. 2006. 3-D target localization using 2-D local displacements of skeletal structures in orthogonal x-ray images for image-guided spinal radiosurgery. *International journal of computer assisted radiology and surgery.* 1:198–200.

Furweger, C., Drexler, C., Kufeld, M., et al. 2010. Patient motion and targeting accuracy in robotic spinal radiosurgery: 260 single-fraction fiducial-free cases. *International journal of radiation oncology, biology, physics.* 78(3):937–945.

Furweger C., Drexler, C., Kufeld, M. et al. 2011. Advances in fiducial-free image-guidance for spinal radiosurgery with CyberKnife—a phantom study. *Journal of applied clinical medical physics.* 12(2):20–28.

Garibaldi, C., Piperno, G., Ferrari, A., et al. 2016. Translational and rotational localization errors in cone-beam CT based image-guided lung stereotactic radiotherapy. *Physica medica.* 32:859–865.

Gayou, O., Parda, D. S., Johnson, M., et al. 2007. Patient dose and image quality from mega-voltage cone beam computed tomography imaging. *Medical physics.* 34:499–506.

Geets, X., Lee, J., Bol, A. 2007. A gradient-based method for segmenting FDG-PET images: Methodology and validation. *European journal of nuclear medicine and molecular imaging.* 34:1427–1438.

Gevaert, T., Verellen, D., Engels, B., et al. 2012. Clinical evaluation of a robotic 6-degree of freedom treatment couch for frameless radiosurgery. *International journal of radiation oncology, biology, physics.* 83(1):467–74.

Ghilezan, M., Yan, D., Martinez, A. 2010. Adaptive radiation therapy for prostate cancer. *Seminars in radiation oncology.* 20:130–7.

Ghilezan, M. J., Jaffray, D. A., Siewerdsen, J. H. 2005. Prostate gland motion assessed with cine-magnetic resonance imaging (cine-MRI). *International journal of radiation oncology, biology, physics.* 62:406–417.

Gill, S., Dang, K., Fox, C., et al. 2014. Seminal vesicle intrafraction motion analysed with cinematic magnetic resonance imaging. *Radiation oncology.* 9:174.

Gill, S. K., Reddy, K., Campbell, N., Chen, C., Pearson, D. 2015. Determination of optimal PTV margin for patients receiving CBCT-guided prostate IMRT: Comparative analysis based on CBCT dose calculation with four different margins. *Journal of applied clinical medical physics.* 16(6):252–262.

Glitzner, M., Crijns, S. P., de Senneville, B. D. 2015. On-line MR imaging for dose validation of abdominal radiotherapy. *Physics in medicine and biology.* 60(22):8869–8883.

Gluhchev, G. 2002. *Random and systematic errors evaluation in radiation therapy.* Proceedings of the 10th Mediterranean Conference on Control and Automation, MED2002, Lisbon, Portugal. 9–12 July.

Goetsch, S. J. 2002. Risk analysis of Leksell Gamma Knife Model C with automatic positioning system. *International journal of radiation oncology, biology, physics.* 52:869–877.

Goetsch, S. J. 2015a. Historic development of stereotactic radiosurgery and stereotactic body radiation therapy. In: Benedict, S. et al. (eds.), *Stereotactic Radiosurgery and Stereotactic Body Radiation Therapy.* London, UK: Routledge, Taylor and Francis Group.

Goetsch, S. J. 2015b. Gamma knife. In: Benedict, S. et al. (eds.), *Stereotactic Radiosurgery and Stereotactic Body Radiation Therapy.* London, UK: Routledge, Taylor and Francis Group.

Goff, P. H., Harrison, L. B., Furhang, E., et al. 2017. 2-D kV orthogonal imaging with fiducial markers is more precise for daily image guided alignments than soft-tissue cone beam computed tomography for prostate radiation therapy. *Advances in radiation oncology.* 2:420–428.

Goyal, S., Kataria, T. 2014. Image guidance in radiation therapy: Techniques and applications. *Radiology research and practice.* 705604:1–10.

Green, O., Rankine, L., Cai, B., et al. 2018. First clinical implementation of real-time, real anatomy tracking and radiation beam control. *Medical physics.* 45(8):3728–3740.

Greto, D., Scoccianti, S., Compagnucci, A., et al. 2016. Gamma knife radiosurgery in the management of single and multiple brain metastases. *Clinical neurology and neurosurgery.* 141:43–47.

Gronborg, C., Vestergaard, A., Hoyer, M., et al. 2015. Intra-fractional bladder motion and margins in adaptive radiotherapy for urinary bladder cancer. *Acta oncologica.* 54:1461–1466.

Grosu, A., Weber, W. 2010. PET for radiation treatment planning of brain tumours. *Radiotherapy & oncology.* 96:325–327.

Guo, L., He, X., He, J. 2016. New delay-decomposing approaches to stability criteria for delayed neural networks. *Radiation oncology.* 11(1):123–129.

Haciislamoglu, E., Colak, F., Canyilmaz E., et al. 2015. Dosimetric comparison of left-sided whole-breast irradiation with 3-DCRT, forward-planned IMRT, inverse-planned IMRT, helical Tomotherapy and volumetric arc therapy. *Physica medica.* 31:360–367.

Hadley, S. W., Balter, J. M., Lam, K. L. 2009. Analysis of couch position tolerance limits to detect mistakes in patient setup. *Journal of applied clinical medical physics.* 10(4):207–219.

Hafeez, S., McDonald, F., Lalondrelle, S., et al. 2017. Clinical outcomes of image guided adaptive hypofractionated weekly radiation therapy for bladder cancer in patients unsuitable for radical treatment. *International journal of radiation oncology, biology, physics.* 98(1):115–122.

Hall, E. J. 2006. Intensity-modulated radiation therapy, protons, and the risk of second cancers. *International journal of radiation oncology, biology, physics.* 65(1):1–7.

Hall, E. J., Wuu, C. S. 2003. Radiation-induced second cancers: The impact of 3D-CRT and IMRT. *International journal of radiation oncology, biology, physics.* 56(1):83–88.

Han, C., Chen, Y. J., Liu, A., et al. 2008. Actual dose variation of parotid glands and spinal cord for nasopharyngeal cancer patients during radiotherapy. *International journal of radiation oncology, biology, physics.* 70:1256–1262.

Han, C., Liu, A., Liang, J., et al. 2018. *ASTRO 2018 Abs.* Dosimetric evaluation of treatment plans for a biology-guided radiotherapy system in treatment of nasopharyngeal cancer. *International journal of radiation oncology, biology, physics.* 102(3):Supplement page e527.

Han, E. Y., Paudel, N., Sung J., et al. 2016. Estimation of the risk of secondary malignancy arising from whole-breast irradiation: Comparison of five radiotherapy modalities, including TomoHDA. *Oncotarget.* 7(16):22960–22969.

Hara, W., Soltys, S. G., Gibbs, I. C. 2007. Cyberknife® robotic radiosurgery system for tumor treatment. *Expert review of anticancer therapy.* 7(11):1507–1515.

Harrison, R. M. 2004. Second cancers following radiotherapy: A suggested common dosimetry framework for therapeutic and concomitant exposures. *British journal of radiology.* 77(924):986–990.

Harrison, R. M., Wilkinson, M., Shemilt, A., et al. 2006. Organ doses from prostate radiotherapy and associated concomitant exposures. *British journal of radiology.* 79(942):487–496.

Hatipoglu, S., Mu, Z., Fu, D., et al. 2007. *Evaluation of a robust fiducial tracking algorithm for image-guided radiosurgery.* Proceedings of SPIE 6509, 65090A.

Hauri, P., Halg, R. A., Besserer, J., et al. 2016. A general model for stray dose calculation of static and intensity-modulated photon radiation. *Medical physics.* 43(4):1955–1968.

Hawkins, M. A., Brock, K. K., Eccles, C. 2006. Assessment of residual error in liver position using kV cone-beam computed tomography for liver cancer high-precision radiation therapy. *International journal of radiation oncology, biology, physics.* 66(2):610–619.

Health and Care Professions Council (HCPC). 2013. *HCPC Standards of Proficiency—Radiographers.* London, UK: HCPC.

Heemsbergen, W. D., Hoogeman, M. S., Witte, M. G., et al. 2007. Increased risk of biochemical and clinical failure for prostate patients with a large rectum at radiotherapy planning: Results from the Dutch trial of 68 GY versus 78 Gy. *International journal of radiation oncology, biology, physics.* 67:1418–1424.

Heijkoop, S., Langerak, T., Quint, S., et al. 2014. Clinical implementation of an online adaptive plan-of-the-day protocol for nonrigid motion management in locally advanced cervical cancer IMRT. *International journal of radiation oncology, biology, physics.* 90(3):673–679.

Heijmink, S. W., Scheenen, T. W., van Lin, E. N. 2009. Changes in prostate shape and volume and their implications for radiotherapy after introduction of endorectal balloon as determined by MRI at 3T. *International journal of radiation oncology, biology, physics.* 73(5):1446–1453.

Held, M., Cremers, F., Sneed, P., et al. 2016. Assessment of image quality and dose calculation accuracy on kV CBCT, MV CBCT, and MV CT images for urgent palliative radiotherapy treatments. *Journal of applied clinical medical physics.* 17(2):279–290.

Hellebust, T. P., Heikkila, I. E., Frykholm, G., et al. 2014. Quality assurance in radiotherapy on a national level; experience from Norway: the KVIST initiative. *Journal of radiotherapy in practice.* 13:35–44.

Helmbrecht, S., Enghardt, W., Fiedler, F., et al. 2016. In-beam PET at clinical proton beams with pile-up rejection. *Zeitschrift für Medizinische Physik (Z Medical physics).* 27:202–217.

Henke, L., Kashani, R., Yang, D. 2016. Simulated online adaptive magnetic resonance-guided stereotactic body radiation therapy for the treatment of oligometastatic disease of the abdomen and central thorax: Characterization of potential advantages. *International journal of radiation oncology, biology, physics.* 96(5):1078–1086.

Hentschel, B., Ochler, W., Straub, D., et al. 2011. Definition of the CTV prostate in CT and MRI by using CT–MRI image fusion in IMRT planning for prostate cancer. *Strahlentherapie und onkologie.* 187:183–190.

Herman, M. G. 2005. Clinical use of electronic portal imaging. *Seminars in radiation oncology.* 15:157–167.

Hideghéty, K., Cserháti, A., Besenyi, Z. 2015. Role of 18FDG-PET/CT in the management and gross tumor volume definition for radiotherapy of head and neck cancer; single institution experiences based on long-term follow-up. *Magyar onkologia.* 59:103–110.

Ho, A. K., Fu, D., Cotrutz, C., et al. 2007. A study of the accuracy of cyberknife spinal radiosurgery using skeletal structure tracking. *Neurosurgery.* 60:147–156.

Hoggarth, M. A., Kuce, J., Syeda, F., et al. 2013. Dual energy imaging using clinical on-board imaging system. *Physics in medicine and biology.* 58:4331–4340.

Hoisak, J. D. P., Pawlicki, T. 2018. The role of optical surface imaging systems in radiation therapy. *Seminars in radiation oncology.* 28(2):185–193.

Hojo, H., Dohmae, T., Hotta, K., et al. 2017. Difference in the relative biological effectiveness and DNA damage repair processes in response to proton beam therapy according to the positions of the spread out Bragg peak. *Radiation oncology.* 12(1):111.

Hong, K., Choi, Y., Jung, J., et al. 2013. A prototype MR insertable brain PET using tileable GAPD arrays. *Medical physics.* 40(4):042503-1–12.

Hoopes, D. J., Dicker, A. P., Eads, N. L., et al. 2015. RO-ILS: Radiation oncology incident learning system: A report from the first year of experience. *Practical radiation oncology.* 5:312–328.

Hricak, H., Brenner, D., Adelstein, S., et al. 2011. Managing radiation use in medical imaging: A multifaceted challenge. *Radiology.* 258:889–905.

Hsu, S-H., Cao, Y., Huang, K., et al. 2013. Investigation of a method for generating synthetic CT models from MRI scans of the head and neck for radiation therapy. *Physics in medicine and biology.* 58(23):8419–8435.

Hu, Y. L., Rankine, L., Green, O. L., et al. 2015. Characterization of the onboard imaging unit for the first clinical magnetic resonance image guided radiation therapy system. *Medical physics.* 42:5828–5837.

Hugo, G., Ruso, M. 2012. Advances in 4-D radiation therapy for managing respiration. *Medical physics.* 22(4):258–271.

Hui, C., Wen, Z., Stemkens, B., et al. 2016. 4D MR imaging using robust internal respiratory signal. *Physics in medicine and biology*. 61(9):3472–3487.

Hui, Z., Zhang, X., Starkschall, G., et al. 2008. Effects of interfractional motion and anatomic changes on proton therapy dose distribution in lung cancer. *International journal of radiation oncology, biology, physics*. 72(5):1385–1395.

Hunt, A., Hansen, V., Oelfke, U., et al. 2018. Adaptive radiotherapy enabled by MRI guidance. *Clinical oncology (Royal College of Radiologists)*. 30:711–719.

Hurkmans, C. W., Remeijer, P., Lebesque, J. V., Mijnheer, B. J. 2002. Set-up verification using portal imaging; review of current clinical practice. *Radiotherapy & oncology*. 58:105–120.

IAEA. 2000. *Safety report series no. 17. Lessons learned from accidental exposures in radiotherapy*. Vienna, Austria: International Atomic Energy Agency.

IAEA. 2007. *Imaging in radiotherapy: Report of the consultant's meeting held in Vienna 15–19 October 2007*. Vienna, Austria: International Atomic Energy Agency.

IAEA. 2013a. *Human health report series no. 7. Record and verify systems for radiation treatment of cancer*. Vienna, Austria: International Atomic Energy Agency.

IAEA. 2013b. *Human health report series no. 257. Roles and responsibilities and education and training requirements for clinically qualified medical physicists*. Vienna, Austria: International Atomic Energy Agency.

IAEA. 2014. *Training Course Series 58: A handbook for the education of radiation therapists (RTTs)*. Vienna, Austria: International Atomic Energy Agency.

IAEA. 2016a. *Application of the risk matrix method to radiotherapy. IAEA TECDOC 1685*, Vienna, Austria: International Atomic Energy Agency.

IAEA. 2016b. *Human health report series no. 31. Accuracy requirements and uncertainties in radiotherapy*. Vienna, Austria: International Atomic Energy Agency.

IAEA, 2018. *International Atomic Energy Agency. Preventing accidental exposure in radiotherapy*. Available at: https://rpop.iaea.org/RPOP/RPoP/Content/AdditionalResources/Training/1_TrainingMaterial/AccidentPreventionRadiotherapy.htm (accessed on 3rd December 2018).

IBA website. *IBA Proton Therapy: Proteus® ONE, Think big, scale smart*. Available at: https://iba-worldwide.com/proton-therapy/proton-therapy-solutions/proteus-one (accessed on 3rd December 2018).

Ilicic, K., Combs, S. E., Schmid, T. E. 2018. New insights in the relative radiobiological effectiveness of proton irradiation. *Radiation oncology*. 13:6.

India 2018. Varian website .*Sterling Cancer Hospital treats first patient in India on Varian Halcyon System*. Available at: https://www.varian.com/news/sterling-cancer-hospital-treats-first-patient-india-varian-halcyon-system (accessed on 3rd December 2018).

Institute of Physics and Engineering in Medicine (IPEM) website .Available at: https://www.ipem.ac.uk/ (accessed on 3rd December 2018).

Institute of Physics and Engineering Medicine (IPEM). 2002. *Medical and Dental Guidance Notes. A Good Practice Guide on all Aspects of Ionising Radiation Protection in the Clinical Environment*. York, UK: IPEM Publications.

International Atomic Energy Agency, IAEA TECDOC 1588. 2008. *Transition from 2-D radiotherapy to 3-D conformal and intensity modulated radiotherapy*. Vienna: International Atomic Energy Agency.

International Commission on Radiation Units and Measurements. 1993. ICRU Report 50. *Prescribing, recording, and reporting photon beam therapy*. Bethesda, MD: ICRU.

International Commission on Radiation Units and Measurements. 1999. ICRU Report 62. *Prescribing, recording, and reporting photon beam therapy*. Supplement to ICRU Report 50. Bethesda, MD: ICRU.

IRMER. 2017. *The Ionising Radiation (Medical Exposure) Regulations 2017. SI 2017/1322*. London, UK: The Stationery Office.

IRR17. 2017. *Ionising Radiations Regulations 2017. SI 2017/1075.* London, UK: The Stationery Office.

Ishikawa, M., Yamaguchi, S., Tanabe, S., et al. 2010. Conceptual design of PET-linac system for molecular-guided radiotherapy. *International journal of radiation oncology, biology, physics.* 78(3)(Suppl):S674.

iSixSigma website. *The deming cycle PDCA.* Available at: https://www.isixsigma.com/dictionary/deming-cycle-pdca/ (accessed on 3rd December 2018).

Jacobs, R., Prabhu, A. V., Monaco, E. A., et al. 2018. Patient perception of gamma knife stereotactic radiosurgery through twitter and Instagram. *Interdisciplinary Neurosurgery.* 13:138–140.

Jadvar, H., Colletti, P. M. 2014. Competitive advantage of PET/MRI. *European journal of radiology.* 83(1):84–94.

Jaffray, D., Carlone, M., Milosevic, M., et al. 2014. A facility for magnetic resonance–guided radiation therapy. *Seminars in radiation oncology.* 24:193–195.

Jaffray, D. A. 2005. Emergent technologies for 3-dimensional image-guided radiation delivery. *Seminars in radiation oncology.* 15(3):208–216.

Jaffray, D. A., Drake, D. G., Moreau, M., et al. 1999. A radiographic and tomographic imaging system integrated into a medical linear accelerator for localization of bone and soft-tissue targets. *International journal of radiation oncology, biology, physics.* 45:773–789.

Jaffray, D. A., Siewerdsen, J. H. 2000. Cone-beam computed tomography with a flat-panel imager: Initial performance characterization. *Medical physics.* 27:1311–1323.

Jan, N., Balik, S., Hugo, G. D., 2014. Interfraction displacement of primary tumor and involved lymph nodes relative to anatomic landmarks in image guided radiation therapy of locally advanced lung cancer. *International journal of radiation oncology, biology, physics.* 88(1):210–215.

Jang, S. Y., Lalonde, R., Ozhasoglu, C., et al. 2016. Dosimetric comparison between cone/iris-based and InCise MLC-based Cyber Knife plans for single and multiple brain metastases. *Journal of applied clinical medical physics.* 17(5):184–199.

Jeraj, R., Mackie, T. R., Balog J., et al. 2004. Radiation characteristics of helical tomotherapy. *Medical physics.* 31(2):396–404.

Jin, J-Y., Yin, F-F., Tenn, S. E., et al. 2008. Use of Brainlab ExacTrac x-ray 6D system in image-guided radiotherapy. *Medical dosimetry.* 33(2):124–134.

Johansson, A., Karlsson, M., Nyholm, T. 2011. CT substitute derived from MRI sequences with ultrashort echo time. *Medical physics.* 38(5):2708–2714.

Johnson, R. P. 2018. Review of medical radiography and tomography with proton beams. *Reports on progress in physics.* 81:016701.

Johnston, A. M. 2006. *Unintended overexposure of patient Lisa Norris during radiotherapy treatment at the Beatson Oncology Centre, Glasgow in January 2006.* Edinburgh, UK: Scottish Executive.

Johnston, A. M. 2015. *Unintended overexposure of a patient during radiotherapy treatment at the Edinburgh Cancer Centre, in September 2015.* Edinburgh, UK: Scottish Executive.

Johnston, H., Hilts, M., Beckham, W., et al. 2008. 3-D ultrasound for prostate localization in radiation therapy: A comparison with implanted markers. *Medical physics.* 35(6):2403–2413.

Jones, D., Suit, H., Akine, Y., et al. 2007. Prescribing, recording and reporting proton beam therapy. *Journal of the ICRU.* 7(2).

Jonsson, J., Karlsson, M., Karlsson, M., et al. 2010. Treatment planning using MRI data: An analysis of the dose calculation accuracy for different treatment regions. *Radiation oncology.* 5:62.

Jung, J. H., Cho, K. H., Kim Y. H., et al. 2012. Effect of jaw size in megavoltage CT on image quality and dose. *Medical physics.* 39(8):4967–4983.

Kaidar-Person, O., Kostich, M., Zagar, T. M., et al. 2016. Helical tomotherapy for bilateral breast cancer: Clinical experience. *The Breast.* 28:79–83.

Kainz, K., Lim, S., Chen, G-P., et al. 2017. PreciseART™ adaptive radiation therapy software: Dose monitoring, re-planning, and delivery verification. Sunnyvale, California: Accuray Inc.

Kamino, Y., Miura, S., Kokubo, M., et al. 2006. Development of a four-dimensional image-guided radiotherapy system with a gimbaled x-ray head. *International journal of radiation oncology, biology, physics.* 66(1):271–278.

Kamino, Y., Miura, S., Kokubo, M., et al. 2007. Development of an ultrasmall C-band linear accelerator guide for a four-dimensional image-guided radiotherapy system with a gimbaled x-ray head. *Medical physics.* 34(5):1797–1808.

Kapanen, M., Tenhunen, M. 2013. T1/T2*-weighted MRI provides clinically relevant pseudo-CT density data for the pelvic bones in MRI-only based radiotherapy treatment planning. *Acta oncologica.* 52(3):612–618.

Kashani, R., Olsen, J., 2018. Magnetic resonance imaging for target delineation and daily treatment modification. *Seminars in radiation oncology.* 28:178–184.

Kassaee, A., Das, I. J., Tochner, Z., et al. 2003. Modification of Gill-Thomas-Cosman frame for extracranial head-and-neck stereotactic radiotherapy. *International journal of radiation oncology, biology, physics.* 57(4):1192–1195.

Kathriarachchi, V., Shang, C., Evans, G., et al. 2016. Dosimetric and radiobiological comparison of Cyber Knife M6™ InCise multileaf collimator over IRIS™ variable collimator in prostate stereotactic body radiation therapy. *Journal of medical physics.* 41:135–143.

Keall, P., Barton, M., Crozier, S. 2014. The Australian magnetic resonance imaging–linac program. *Seminars in radiation oncology.* 24:203–206.

Keall, P. J., Mageras, G. S., Balter, J. M., et al. 2006. The management of respiratory motion in radiation oncology. *Medical physics.* 44:3874–3900.

Keller, H., Glass, M., Hinderer, R., et al. 2002. Monte Carlo study of a highly efficient gas ionization detector for megavoltage imaging and image-guided radiotherapy. *Medical physics.* 29(2):165–175.

Kennedy, C., Ling, C., Scheurman, R., et al. 2017. Performance of a new EPID panel and opportunities for a fast MV-CBCT acquisition. *Radiotherapy & oncology.* 123(Suppl 1):S956–S957.

Khoo, V., Joon, D. 2006. New developments in MRI for target volume delineation in radiotherapy. *British journal of radiology.* 79:S2–S15.

Kilby, W., Dooley, J. R., Kuduvalli, G., et al. 2010. CyberKnife® Robotic Radiosurgery System in 2010. *Technology in cancer research & treatment.* 9(5):433–452.

Kim, J., Jang, H. S., Kim, Y. S., et al. 2017a. Comparison of spinal stereotactic body radiotherapy (SBRT) planning techniques: Intensity-modulated radiation therapy, modulated arc therapy and helical Tomotherapy. *Medical dosimetry.* 42:210–215.

Kim, J. I., Lee, H., Wu, H. G. 2017b. Development of patient-controlled respiratory gating system based on visual guidance for magnetic-resonance image-guided radiation therapy. *Medical physics.* 44(9):4838–4846.

Kim, N., Lee, H., Kim, J. S., et al. 2017c. Clinical outcomes of multileaf collimator-based CyberKnife for spine stereotactic body radiation therapy. *British journal of radiology.* 90:20170523.

Kinhikar, R. A., Jamema, S. V., Reenadevi, et al. 2009a. Dosimetric validation of first helical tomotherapy Hi-Art II machine in India. *Journal of medical physics.* 34(1):23–30.

Kinhikar, R. A., Master, Z., Dhote, D. S., Deshpande, D. D. 2009b. Initial dosimetric experience with mega voltage computed tomography detectors and estimation of pre and post-repair dosimetric parameters of a first Helical Hi-Art II tomotherapy machine in India. *Journal of medical physics.* 34(2):73–79.

Kirby, M., Kane, B., Williams, P. 1995. Clinical applications of composite and realtime megavoltage imaging. *Clinical oncology journal (Royal College of Radiologists).* 7(5):308–316.

Kirby, M. C. 1996. The consequences of fixed pattern noise and image movement on electronic portal images. *Physics in medicine & biology.* 41:2345–2356.

Kirby, M. C. 2015. Teaching radiotherapy physics concepts using simulation: Experience with student radiographers in Liverpool, UK. *Medical physics international journal.* 3(2):87–93. Available at: http://www.mpijournal.org/MPI-v03i02.aspx (accessed on 3rd December 2018).

Kirby, M. C., Davies, J. 2010. *Buyer's Guide: Radiotherapy external beam record and verify/oncology management systems.* NHS Centre for Evidence Based Purchasing (CEP10046). London, UK: Crown.

Kirby, M. C., Glendinning, A. G. 2006. Developments in electronic portal imaging systems. *British journal of radiology.* 79:S50–S65.

Kirby, M. C., Kane, B., Williams, P.C. 1995. Clinical applications of composite and realtime megavoltage imaging. *Clinical oncology (Royal College of Radiologists).* 7(5):308–316.

Kirby, M. C., Carpenter, D., Lawrence, G., et al. (eds.) 2006a. *Guidance for the commissioning and quality assurance of a networked radiotherapy department: IPEM Report 93.* York, UK: Institute of Physics and Engineering in Medicine.

Kirby, M. C., Pennington, H., Al-Samarraie, F., et al. 2014. Clinical technology in 21st century radiotherapy education—towards greater alignment with clinical competencies. E-poster presentation at ESTRO 33, 4–8 April 2014, Vienna, Austria. *Radiotherapy oncology.* 111(Suppl 1):738.

Kirby, M. C., Ryde, S., Hall, C. (eds). 2006b. *Acceptance testing and commissioning of linear accelerators: IPEM Report 94.* York, UK: Institute of Physics and Engineering in Medicine.

Kirchner, J., Deuschl, C., Schweiger, B., et al. 2017. Imaging children suffering from lymphoma: An evaluation of different 18F-FDG PET/MRI protocols compared to whole-body DW-MRI. *European journal of nuclear medicine and molecular imaging.* 44:1742–1750.

Kirk, D., Kirby, M. C. 2018. *Understanding range uncertainties in proton beam therapy for paranasal cancer for the new UK service—a critical review of the evidence base.* Presented at MPEC 2018 IPEM Medical Physics and Engineering Conference, 18–20 September 2018, York, UK.

Kirkby, C., Stanescu, T., Rathee, S., et al. 2008. Patient dosimetry for hybrid MRI-radiotherapy systems. *Medical physics.* 35:1019–1027.

Kirkpatric, J. P., Wang, Z., Sampson, J. H., et al. 2015. Defining the optimal planning target volume in image-guided stereotactic radiosurgery of brain metastases: Results of a randomized trial. *International journal of radiation oncology, biology, physics.* 91(1):100–108.

Klein, E. E., Hanley, J., Bayouth, J., et al. 2009. Quality assurance of medical accelerators. *Medical physics.* 36(9):4197–4212.

Knoos, T. 2017. Lessons learnt from past incidents and accidents in radiation oncology. *Clinical oncology (Royal College of Radiologists).* 29:557–561.

Knopf, A-C., Lomax, A. 2013. In vivo proton range verification: A review. *Physics in medicine and biology.* 58:R131–R160.

Kocher, M., Wittig, A., Piroth, M. D., et al. 2014. Stereotactic radiosurgery for treatment of brain metastases. *Strahlentherapie und Onkologie.* 190(6):521–532.

Kong, V., Taylor, A., Chung, P., et al. 2018b. Comparison of 3 image-guided adaptive strategies for bladder locoregional radiotherapy. *Medical dosimetry.* (In press)

Kong, V. C., Taylor, A., Chung, P., et al. 2018a. Evaluation of resource burden for bladder adaptive strategies: A timing study. *Journal of medical imaging and radiation oncology.* (In press)

Kontaxis, C., Bol, G., Lagendijk, J., et al. 2015. A new methodology for inter- and intrafraction plan adaptation for the MR-linac. *Physics in medicine and biology.* 60(19):7485.

Kontaxis, C., Bol, G. H, Stemkens, B. 2017. Towards fast online intrafraction replanning for free-breathing stereotactic body radiation therapy with the MR-linac. *Physics in medicine and biology.* 62(18):7233–7248.

Korhonen, J., Kapanen, M., Keyrilainen, J., et al. 2014. A dual model HU conversion from MRI intensity values within and outside of bone segment for MRI-based radiotherapy treatment planning of prostate cancer. *Medical physics.* 41(1):011704.

Korreman, S., Rasch, C., McNair, H., et al. 2010. The European Society of Therapeutic Radiology and Oncology–European Institute of Radiotherapy (ESTRO–EIR) report on 3-D CT-based in-room image guidance systems: A practical and technical review and guide. *Radiotherapy & oncology.* 94:129–144.

Kothary, N., Heit, J. J., Louie, J. D. 2009. Safety and efficacy of percutaneous fiducial marker implantation for image-guided radiation therapy. *Journal of vascular and interventional radiology.* 20:235–239.

Kraan, A. C., van de Water, S., Teguh, D. N., et al. 2013. Dose uncertainties in IMPT for oropharyngeal cancer in the presence of anatomical, range, and setup errors. *International journal of radiation oncology, biology, physics.* 87:888–896.

Krengli, M., Loi, G., Pisani, C., et al. 2016. Three-dimensional surface and ultrasound imaging for daily IGRT of prostate cancer. *Radiation oncology.* 11(1):159–166.

Kubo, H., Len, P., Minohara, S. et al. 2000. Breathing-synchronized radiotherapy program at the University of California Davis Cancer Center. *Medical physics.* 27(2):346–353.

Kuo, J. S., Yu, C., Giannotta, S. L., et al. 2004. The Leksell Gamma Knife Model U versus Model C: A quantitative comparison of radiosurgical treatment parameters. *Neurosurgery.* 55:168–173.

Kupelian, P., Langen, K., Willoughby, T., et al. 2008a. Image-guided radiotherapy for localized prostate cancer: Treating a moving target. *Seminars in radiation oncology.* 18:58–66.

Kupelian, P., Sonke, J. J. 2014. Magnetic resonance–guided adaptive radiotherapy: A solution to the future. *Seminars in radiation oncology.* 24:227–232.

Kupelian, P. A., Langen, K. M., Zeidan, O. A., et al. 2006. Daily variations in delivered doses in patients treated with radiotherapy for localized prostate cancer. *International journal of radiation oncology, biology, physics.* 66:876–882.

Kupelian, P. A., Lee, C., Langen, K. M. 2008b. Evaluation of image-guidance strategies in the treatment of localized prostate cancer. *International journal of radiation oncology, biology, physics.* 70(4):1151–1157.

Kwint, M., Conijn, S., Schaake, E., et al. 2014. Intra thoracic anatomical changes in lung cancer patients during the course of radiotherapy. *Radiotherapy & oncology.* 113:392–397.

Lagendijk, J., Raaymakers, B., van Vulpen, M. 2014. The magnetic resonance imaging–linac system. *Seminars in radiation oncology.* 24:207–209.

Lagendijk, J., van Vulpen, M., and Raaymakers, B. 2016. The development of the MRI linac system for online MRI-guided radiotherapy: A clinical update. *Journal of internal medicine.* 280:203–208.

Landry, G., Nijhuuis, R., Dedes, G., et al. 2015. Investigating CT to CBCT image registration for head and neck proton therapy as a tool for daily dose recalculation. *Medical physics.* 42(3):1354–1366.

Langen, K. M., Papanikolaou, N., Balog J., et al. 2010. QA for helical tomotherapy: Report of the AAPM Task Group 148. *Medical physics.* 37(9):4817–4853.

Langendijk, J. A., Boersma, L., Rasch, C., et al. 2018. Clinical trial strategies to compare protons with photons. *Seminars in radiation oncology.* 28(2):79–87.

Langmack, K. A. 2001. Portal imaging. *British journal of radiology.* 74:789–804.

Langner, U., Eley, J., Dong, L. et al. 2017. Comparison of multi-institutional Varian ProBeam pencil beam scanning proton beam commissioning data. *Journal of applied clinical medical physics.* 18(3):96–107.

Lau, S., Patel, K., Kim, T., et al. 2017. Clinical efficacy and safety of surface imaging guided radiosurgery (SIG-RS) in the treatment of benign skull base tumors. *Journal of neuro-oncology.* 132(2):307–312.

Leclerc, M., Lartigau, E., Lacornerie, T. 2015. Primary tumor delineation based on (18)FDG PET for locally advanced head and neck cancer treated by chemo-radiotherapy. *Radiotherapy & oncology.* 116:87–93.

Lee, C., Langen, K. M., Lu, W., et al. 2008. Assessment of parotid gland dose changes during head and neck cancer radiotherapy using daily megavoltage computed tomography and deformable image registration. *International journal of radiation oncology, biology, physics.* 70:1563–1571.

Lee, S. W., Jin, J. Y., Guan, H., et al. 2008. Clinical assessment and characterization of a dual tube kilovoltage x-ray localization system in the radiotherapy treatment room. *Journal of applied clinical medical physics*. 9:2318.

Leer, J., McKenzie, A., Scalliet, P., et al. 1998. *Practical guidelines for the implementation of a quality system in radiotherapy*. Brussels, Belgium: ESTRO.

Lehman, J., Perks, J., Semon, S., et al. 2007. Commissioning experience with cone-beam computed tomography for image-guided radiation therapy. *Journal of applied clinical medical physics*. 8(3):21–36.

Leroy, R., Benahmed, N., Hulstaert, F., et al. 2015. Proton therapy in children: A systematic review of clinical effectiveness in 15 pediatric cancers. *International journal of radiation oncology, biology, physics*. 95(1):267–278.

Leuven 2017. *UZ Leuven treats first patient in Europe on Varian Halcyon Cancer Treatment System*. Available at: https://www.prnewswire.com/news-releases/uz-leuven-treats-first-patient-in-europe-on-varian-halcyon-cancer-treatment-system-300537606.html (accessed on 3rd December 2018).

Li, G., Ballangrud, A., Chan, M., et al. 2015. Clinical experience with two frameless stereotactic radiosurgery (fSRS) systems using optical surface imaging for motion monitoring. *Journal of applied clinical medical physics*. 16(4):149–162.

Li, G., Ballangurd, A., Kuo, L. C., et al. 2011. Motion monitoring for cranial frameless stereotactic radiosurgery using video-based three-dimensional optical surface imaging. *Medical physics*. 38(7):3981–3994.

Li, H., Chen, H. C., Dolly, S., et al. 2016. An integrated model-driven method for in-treatment upper airway motion tracking using cine MRI in head and neck radiation therapy. *Medical physics*. 43:4700–4710.

Li, H., Zhu, R., Zhang, L., et al. 2008. Comparison of 2-D radiographic images and 3-D cone beam computed tomography for positioning head-and-neck radiotherapy patients. *International journal of radiation oncology, biology, physics*. 71(3):916–925.

Li, M., Hegemann, N-S., Manapov, F., et al. 2017. Prefraction displacement and intrafraction drift of the prostate due to perineal ultrasound probe pressure. *Strahlentherapie und onkologie*. 193:459–465.

Li, W., Cho, Y. B., Ansell, S., et al. 2016. The use of cone-beam computed tomography for image guided gamma knife stereotactic radiosurgery: Initial clinical evaluation. *International journal of radiation oncology, biology, physics*. 96(1):214–220.

Li, Y., Netherton, T., Nitsch, P. L., et al. 2018. Normal tissue doses from MV image-guided radiation therapy (IGRT) using orthogonal MV and MV-CBCT. *Journal of applied clinical medical physics*. 19(3):52–57.

Liang, J., Liu, A., Han, C., et al. 2018. *ASTRO 2018 Abs*. A dosimetric study to assess the feasibility of prototype treatment planning software for a new biology-guided radiotherapy system. *International journal of radiation oncology, biology, physics*. 102(3):Supplement page e477.

Liao, Z., Gandhi, S., Lin, S., et al. 2008. Does proton therapy offer demonstrable clinical advantages for treating thoracic tumors? *Seminars in radiation oncology*. 28:114–124.

Lim, K., Small, W., Portelance, L., et al 2011. Consensus guidelines for delineation of clinical target volume for intensity-modulated pelvic radiotherapy for the definitive treatment of cervix cancer. *International journal of radiation oncology, biology, physics*. 79(2):348–355.

Lin, P-J., Beck, T. J., Borras, C. et al. 1993. *Specification and acceptance testing of computed tomography scanners: Report of Task Group 2 Diagnostic X-Ray Imaging Committee*. College Park, MD, USA: American Association of Physicists in Medicine.

Lin, Z., Mechalakos, J., Nehmeh, S. 2007. The influence of changes in tumor hypoxia on dose-painting treatment plans based on 18F-FMISO positron emission tomography. *International journal of radiation oncology, biology, physics*. 70:1219–1228.

Lindsay, K. A., Wheldon, E. G., Deehan, C., et al. 2001. Radiation carcinogenesis modelling for risk of treatment-related second tumours following radiotherapy. *British journal of radiology.* 74(882):529–536.

Liney, G., Dong, B., Begg, J., et al. 2016. Technical Note: Experimental results from a prototype high-field inline MRI-linac. *Medical physics.* 43(9):5188–5194.

Liney, G., Moerland, M. 2014. Magnetic resonance imaging acquisition techniques for radiotherapy planning. *Seminars in radiation oncology.* 24:160–168.

Liney, G., Whelan, B., Obron, B., et al. 2018. MIR-linear accelerator radiotherapy systems. *Clinical oncology (Royal College of Radiologists).* 30:686–691.

Litzenberg, D. W., Balter, J. M., Lam, K. L., et al. 2005. Retrospective analysis of prostate cancer patients with implanted gold markers using off-line and adaptive therapy protocols. *International journal of radiation oncology, biology, physics.* 63(1):123–133.

Liu, L., Cao, Y., Fessler, J., et al. 2016. A female pelvic bone shape model for air/bone separation in support of synthetic CT generation for radiation therapy. *Physics in medicine and biology.* 61(1):169–182.

Liu, W., Liao, Z., Schild, S., et al. 2015. Impact of respiratory motion on worst-case scenario optimized intensity modulated proton therapy for lung cancers *Practical radiation oncology.* 5:e77–e78.

Liu, Y., Yin, F. F., Chang, Z., et al. 2014. Investigation of sagittal image acquisition for 4D-MRI with body area as respiratory surrogate. *Medical physics.* 41(10):101902.

Lopes, P. C., Clementel, E., Crespo, P., et al. 2015. Time-resolved imaging of prompt-gamma rays for proton range verification using a knife-edge slit camera based on digital photon counters. *Physics in medicine and biology.* 60:6063–6085.

Lovelock, D., Messineo, A., Cox, B., et al. 2015. Continuous monitoring and intrafraction target position correction during treatment improves target coverage for patients undergoing SBRT prostate therapy. *International journal of radiation oncology biology physics.* 91(3):588–594.

Lovelock, D. M., Zatcky, J., Goodman, K., Yamada, Y. 2014. The effectiveness of a pneumatic compression belt in reducing respiratory motion of abdominal tumors in patients undergoing stereotactic body radiotherapy. *Technology in cancer research and treatment.* 13(3):259–267.

Lowe, C., Baker, A., Kirby, M. C. 2016. *VMAT with image guided verification for rectal cancer patients; a service evaluation to develop and justify current CBCT protocol.* Poster presented at UKRCO 2016 UK Radiation Oncology Congress, 6–7 June 2016, Liverpool, UK.

Lu, W., Chen, M., Chen, Q., et al. 2008. Adaptive fractionation therapy: I. Basic concept and strategy. *Physics in medicine and biology.* 53:5495–5511.

Ludbrook, J. J. S., Greer, P. B., Blood, P., et al. 2005. Correction of systematic setup errors in prostate radiation therapy: How many images to perform? *Medical dosimetry.* 30(2):76–84.

Ma, C., Lomax, E., (eds.) 2012. *Proton and Carbon Ion Therapy.* Boca Raton, FL, USA: CRC Press, Taylor and Francis Group.

Ma, J., Chang, Z., Wang, Z., et al. 2009. ExacTrac x-ray 6 degree-of-freedom image-guidance for intracranial non-invasive stereotactic radiotherapy: Comparison with kilo-voltage cone-beam CT. *Radiotherapy & oncology.* 93:602–608.

Mackie, T. R. 2006. History of tomotherapy. *Physics in medicine and biology.* 51:R427–R453.

Magome, T., Haga, A., Takahashi Y., et al. 2016. Fast megavoltage computed tomography: A rapid imaging method for total body or marrow irradiation in helical tomotherapy. *International journal of radiation oncology, biology, physics.* 96(3):688–695.

Malapure, S., Das, K., Kumar, R. 2016. PET/Computed tomography in breast cancer: Can it aid in developing a personalized treatment design? *PET clinics.* 11:297–303.

Malicki, J., Bly, R., Bulot, M., et al. 2014. Patient safety in external beam radiotherapy—guidelines on risk assessment and analysis of adverse error-events and near-misses: Introducing the ACCIRAD project. *Radiotherapy & oncology.* 112:194–198.

Malicki, J., Bly, R., Bulot, M., et al. 2018. Patient safety in external beam radiotherapy, results of the ACCIRAD project: Recommendations for radiotherapy institutions and national authorities on assessing risks and analysing adverse error-events and near misses. *Radiotherapy & oncology* 127(2):164–170.

Manger, R., Paxton, A., Pawlicki, T., et al. 2015. Failure mode and effects analysis and fault tree analysis of surface image guided cranial radiosurgery. *Medical physics.* 42(5):2449–2461.

Mao, W., Liu, C., Snyder, K. et al. 2017. *A Novel Iterative Reconstruction Algorithm for Improving CBCT Image Quality.* AAPM Annual Meeting, 2017.

Martins, L., Couto, J. G., Barbosa, B. 2016. Use of planar kV vs. CBCT in evaluation of setup errors in oesophagus carcinoma radiotherapy. *Reports of practical oncology and radiotherapy.* 21:57–62.

Martisikova, M., Gehrke, T., Berke, S., et al. 2018. Helium ion beam imaging for image guided ion radiotherapy. *Radiation oncology.* 13:109.

Martyn, M., O'Shea, T., Harris, E., et al. 2017. A Monte Carlo study of the effect of an ultrasound transducer on surface dose during intrafraction motion imaging for external beam radiation therapy. *Medical physics.* 44(10):5020–5033.

Matsuo, Y., Ueki, N., Takayama, K., et al. 2014. Evaluation of dynamic tumour tracking radiotherapy with real-time monitoring for lung tumours using a gimbal mounted linac. *Radiotherapy & oncology.* 112:360–364.

Matsuura, T., Nishimura, Y., Nakamatsu, K. 2017. Clinical outcomes of IMRT planned with or without PET/CT. *International journal of clinical oncology.* 22(1):52–58.

Mayles, P., Nahum, A., Rosenwald, J-C. 2007. *Handbook of Radiotherapy Physics: Theory and Practice.* New York, USA: Taylor and Francis Group.

McCullough, K. T., James, J. A., Cetnar, A. J., et al. 2015. Site-specific tolerance tables and indexing device to improve patient setup reproducibility. *Journal of applied clinical medical physics.* 16(3):378–388.

McGary, J. E., Teh, B. S., Butler, B., Grant III, W. 2002. Prostate immobilization using a rectal balloon. *Journal of applied medical physics.* 3(1):6–11.

McKenzie, A. L., van Herk, M., Mijnheer, B. 2000. The width of margins in radiotherapy treatment plans. *Physics in medicine and biology.* 45:3331–3342.

McNair, H. A., Hafeez, S., Taylor, H., et al. 2015. Radiographer-led plan selection for bladder cancer radiotherapy: Initiating a training programme and maintaining competency. *British journal of radiology.* 88:20140690.

McPartlin, A. J., Li, X. A., Kershaw, L. E., et al. 2016. MRI-guided prostate adaptive radiotherapy—a systematic review. *Radiotherapy & oncology.* 119(3):371–380.

McTyre, E., Helis, C. A., Farris, M., et al. 2017. Emerging indications for fractionated gamma knife radiosurgery. *Neurosurgery.* 80(2):210–216.

Medphoton website. *MedPhoton ImagingRing System.* Available at: https://www.medphoton.at/products/imaging-ring/ (accessed on 3rd December 2018).

Meeks, S., Bova, F., Wagner, T., et al. 2000. Image localization for frameless stereotactic radiotherapy. *International journal of radiation oncology, biology, physics.* 46(5):1291–1299.

Meeks, S, Willoughby, T., Langen, K., et al. 2012. Optical and remote monitoring IGRT. In: Bourland, D. (ed.), *Image Guided Radiation Therapy.* Boca Raton, FL, USA: CRC Press, Taylor and Francis Group.

Mege, J-P., Wenzhao, S., Veres, A., et al. 2016. Evaluation of MVCT imaging dose levels during helical IGRT: Comparison between ion chamber, TLD and EBT3 films. *Journal of applied clinical medical physics.* 17(1):143–157.

Mehta, S., Gajjar, S., Padgett, K., et al. 2018. Daily tracking of glioblastoma resection cavity, cerebral edema, and tumor volume with MRI-guided radiation therapy. *Cureus.* 10(3):e2346–e2355.

Menard, C., van der Heide, U. 2014. Introduction: Systems for magnetic resonance image guided radiation therapy. *Seminars in radiation oncology.* 24:192.

Mendenhall, N. P., Li, Z. 2012. Proton therapy. In: Levitt, S. H., et al. (eds.) *Technical Basis of Radiation Therapy.* 5th ed. Berlin, Germany: Springer-Verlag.

Menke, M., Hirschfeld, F., Mack, T., et al. 1994. Photogrammetric accuracy measurements of head holder systems used for fractionated radiotherapy. *International journal of radiation oncology biology physics.* 29(5):1147–1155.

Merchant, T. E. 2013. Clinical controversies: Proton therapy for pediatric tumors. *Seminars in radiation oncology.* 23:97–108.

Metcalfe, P. E., Kron, T., Hoban, P. 2007. *The Physics of Radiotherapy X-rays and Electrons.* 2nd ed. Madison, WI: Medical Physics Publishing.

Mevion website. *Mevion Proton Therapy: Mevion S250 Proton therapy system.* Available at: http://www.mevion.com/products/mevion-s250-proton-therapy-system (accessed on 3rd December 2018).

Meyer, P., Le Pennec, F., Hui, S. K., et al. 2017. Megavoltage 2-D topographic imaging: An attractive alternative to megavoltage CT for the localization of breast cancer patients treated with TomoDirect. *Physica medica.* 39:33–38.

MHI website. 2012. *Mitsubishi Heavy Industries Technical Review.* Vol 49 (1). Available at: https://www.mhi.co.jp/technology/review/pdf/e491/e491044.pdf (accessed on 3rd December 2018).

Michiels, S., Poels, K., Crijns, W., et al. 2018. Volumetric modulated arc therapy of head-and-neck cancer on a fast-rotating O-ring linac: Plan quality and delivery time comparison with a C-arm linac. *Radiotherapy & oncology.* 128(3):479–484.

Mihailidis, D., Brady, L., Anamalayil, S., et al. 2017b. Rapid IMRT delivery for head and neck (H&N) with a prototype jawless MLC system and a novel MV-CBCT panel. *International journal of radiation oncology, biology, physics.* 99(2):S230–S231.

Mihailidis, D., Schuerman, R., Kennedy, C., et al. 2017a. AAPM TG-119 benchmarking of a novel jawless dual level MLC collimation system. *Radiotherapy & oncology.* 123(Suppl 1):S953–S954.

Mijnheer, B., Beddar, S., Izewska, J. et al. 2013. In vivo dosimetry in external beam radiotherapy. *Medical physics.* 40(7):070903-1–19.

Mijnheer, B., Olszewska, A., Fiorino, C., et al. *Quality assurance of treatment planning systems—practical examples for non-IMRT photon beams.* Brussels, Belgium: ESTRO, 2004.

Miller, J. A., Kotecha, R., Barnett, G., et al. 2016. *Quality of life following Gamma Knife radiosurgery for single and multiple brain metastases.* Proceedings of the 98th Annual Meeting of the American Radium Society (www.americanradiumsociety.org), 15 April 2016.

Milosevic, M., Angers, C., Liszewski, B., et al. 2016. The Canadian national system for incident reporting in radiation treatment (NSIR-RT) taxonomy. *Practical radiation oncology.* 6:334–341.

Min, C., Zhu, X., Winey, B., et al. 2013. Clinical application of in-room PET for in vivo treatment monitoring in proton radiotherapy. *International journal of radiation oncology, biology, physics.* 86(1):183–189.

Miura, H., Ozawa, S., Hayata, M., et al. 2017. Evaluation of cone-beam computed tomography image quality assurance for Vero4DRT system. *Reports of practical oncology and radiotherapy.* 22:258–263.

Miyabe, Y., Sawada, A., Takayama, K., et al. 2011. Positioning accuracy of a new image-guided radiotherapy system. *Medical physics.* 38(5):2535–2541.

Moghanaki, D., Turkbey, B., Vapiwala, N., et al. 2017. Advances in prostate cancer magnetic resonance imaging and positron emission tomography-computed tomography for staging and radiotherapy treatment planning. *Seminars in radiation oncology.* 27:21–33.

Mohan, D., Kupelian, P., Willoughby, T. 2000. Short-course intensity-modulated radiotherapy for localized prostate cancer with daily transabdominal ultrasound localization of the prostate gland. *International journal of radiation oncology biology physics.* 46(3):575–580.

Mohan, R., Das, I., Ling, C. 2017. Empowering intensity modulated proton therapy through physics and technology: An overview. *International journal of radiation oncology, biology, physics.* 99(2):304–316.

Mohan, R., Grosshans, D. 2017. Proton therapy—present and future. *Advanced drug delivery reviews.* 109:26–44.

Moller, D. S., Khalil, A. A., Knap, M. M., et al. 2014. Adaptive radiotherapy of lung cancer patients with pleural effusion or atelectasis. *Radiotherapy & oncology.* 110:517–522.

Molloy, J. 2012. Ultrasound guided radiation therapy. In: Bourland, D. (ed.), *Image Guided Radiation Therapy.* Boca Raton, FL, USA: CRC Press, Taylor and Francis Group.

Molloy, J. A., Chan, G., Markovic, A., et al. 2011. Quality assurance of US-guided external beam radiotherapy for prostate cancer: Report of AAPM Task Group 154. *Medical physics.* 38:857–871.

Montgomery, L., Fava, P., Freeman, C. R., et al. 2018. Development and implementation of a radiation therapy incident learning system compatible with local workflow and a national taxonomy. *Journal of applied clinical medical physics.* 19(1):259–270.

Moore, K. L., Palaniswaamy, G., White, B., et al. 2010. Fast, low-dose patient localization on TomoTherapy via topogram registration. *Medical physics.* 37(8):4068–4077.

Moreau, J., Biau, J., Achard, J-L., et al. 2017. Intraprostatic fiducials compared with bony anatomy and skin marks for image-guided radiation therapy of prostate cancer. *Cureus.* 9(10):e1769.

Morr, J., DiPetrillo, T., Tsai, J., et al. 2002. Implementation and utility of a daily ultrasound-based localization system with intensity-modulated radiotherapy for prostate cancer. *International journal of radiation oncology biology physics.* 53(5):1124–1129.

Morrow, N. V., Lawton, C. A., Qi, X. S., et al. 2012. Impact of computed tomography image quality on image-guided radiation therapy based on soft tissue registration. *International journal of radiation oncology, biology, physics.* 82(5):e733–e738.

Moseley, D. J., White, E. A., Wiltshire, K. L., et al. 2007. Comparison of localization performance with implanted fiducial markers and cone-beam computed tomography for on-line image-guided radiotherapy of the prostate. *International journal of radiation oncology, biology, physics.* 67(3):942–953.

Moteabbed, M., Espana, S., Paganetti, H. 2011. Monte Carlo patient study on the comparison of prompt gamma and PET imaging for range verification in proton therapy. *Physics in medicine and biology.* 56:1063–1082.

Mu, Z., Fu, D., Kuduvalli, G. 2006. Multiple fiducial identification using 54. The hidden Markov model in image guided radiosurgery. In: *Computer Vision and Pattern Recognition IEEE Computer Society.* Washington, DC.

Muacevic, A., Drexler, C. Kufeld, M., et al. 2009. Fiducial-free real-time image-guided robotic radiosurgery for tumors of the sacrum/pelvis. *Radiotherapy & oncology.* 93:37–44.

Mukumoto, N., Nakamura, M., Sawada, A., et al. 2012. Positional accuracy of novel x-ray-image-based dynamic tumor-tracking irradiation using a gimbaled MV x-ray head of a Vero4DRT (MHI-TM2000). *Medical physics.* 39(10):6287–96.

Mukumoto, N., Nakamura, M., Yamada, M., et al. 2014. Intrafractional tracking accuracy in infrared marker-based hybrid dynamic tumour-tracking irradiation with a gimballed linac. *Radiotherapy & oncology.* 111:301–305.

Munro, P., Kirby, M. 2004. *The physics and technology of portal imaging.* Proceedings of the 8th international workshop on electronic portal imaging, 29 June–1 July 2004, Brighton, UK.

Murphy, M., Balter, J. M., Balter, S., et al. 2007. The management of imaging dose during image-guided radiotherapy: Report of the AAPM Task Group 75. *Medical physics.* 34:4041–4063.

Murthy, V., Masodkar, R., Kalyani, N., et al. 2016. Clinical outcomes with dose-escalated adaptive radiation therapy for urinary bladder cancer: A prospective study. *International journal of radiation oncology, biology, physics.* 94(1):60–66.

Mutic, S., Dempsey, J. 2014. The ViewRay System: Magnetic resonance–guided and controlled radiotherapy. *Seminars in radiation oncology*. 24:196–199.

Nagata, Y., Nishidai, T., Abe, M., et al. 1990. CT simulator: A new 3-D planning and simulating system for radiotherapy: Part 2. Clinical application. *International journal of radiation oncology, biology, physics*. 18:505–513.

Nakamura, M., Sawada, A., Ishihara, Y., et al. 2010. Dosimetric characterization of a multileaf collimator for a new four-dimensional image-guided radiotherapy system with a gimbaled x-ray head. *Medical physics*. 37(9):4684–4691.

Naoi, Y., Kunishima, N., Yamamoto, K., et al. 2014. A planning target volume margin formula for hypofractionated intracranial stereotactic radiotherapy under cone beam CT image guidance with a six-degrees-of-freedom robotic couch and a mouthpiece-assisted mask system: A preliminary study. *British journal of radiology*. 87:20140240

National Cancer Peer Review Programme (NCAT). *Manual for Cancer Services—Radiotherapy Measures Version 5.0.* 2013. Available at: https://www.cquins.nhs.uk/download.php?d=resources/measures/Radiotherapy_April2013.pdf (accessed on 3rd December 2018).

National Health Service. 2003. *Health and Social Care (Community Health and Standards) Act 2003.* London, UK: The Stationary Office. Available at: http://www.legislation.gov.uk/ukpga/2003/43/pdfs/ukpga_20030043_en.pdf (accessed on 3rd December 2018).

National Health Service. 2017. *Quality Surveillance Groups—National Guidance.* Available at: https://www.england.nhs.uk/publication/quality-surveillance-groups-national-guidance/ (accessed on 3rd December 2018).

National Radiotherapy Implementation Group (NRIG). 2012. *Image guided radiotherapy—guidance for implementation and use.* London, UK: National Cancer Action Team.

National School of Healthcare Science (NSHCS). 2014. *Scientist Training Programme: MSc in Clinical science Curriculum.* Available at: https://www.nshcs.hee.nhs.uk (accessed 3rd December 2018).

Netherton, T., Li, Y., Nitsch, P., et al. 2017. Efficiency and efficacy of intensity modulated treatments on a novel linear accelerator. *International journal of radiation oncology, biology, physics*. 99(2):E703.

Nijkamp, J., Pos, F. J., Nuver, T. T., et al. 2008. Adaptive radiotherapy for prostate cancer using kilovoltage cone-beam computed tomography: First clinical results. *International journal of radiation oncology, biology, physics*. 70:75–82.

Nilsson, P., Ceberg, C., Kjellen, E. 2015. A template for writing radiotherapy protocols. *Acta Oncologica*. 54:275–279.

Nitsch, P., Li, Y., Netherton, T., et al. 2017. Radiotherapy treatment using a prototype MLC design. *Radiotherapy & oncology*. 123(Suppl 1):S846–S847.

Nobah, A., Aldelaijan, S., Devic S., et al. 2014. Radiochromic film based dosimetry of image-guidance procedures on different radiotherapy modalities. *Journal of applied clinical medical physics*. 15(6):229–239.

Noel, C. E., Parikh, P. J., Spencer, C. R. 2015. Comparison of onboard low-field magnetic resonance imaging versus onboard computed tomography for anatomy visualization in radiotherapy. *Acta oncologica*. 54(9):1474–1482.

Nyflot, M. J., Zeng, J., Kusano, A. S., et al. 2015. Metrics of success: Measuring impact of a department near-miss incident learning system. *Practical radiation oncology*. 5(5):334–341.

Nyholm, T., Jonsson, J. 2014. Counterpoint: Opportunities and challenges of a magnetic resonance imaging–only radiotherapy work flow. *Seminars in radiation oncology*. 24:175–180.

Oborn, B. M., Dowdell, S., Metcalfe, P. E., et al. 2017. Future of medical physics: Real-time MRI-guided proton therapy. *Medical physics*. 44(8):e77–e90.

Oehler, C., Lang, S., Dimmerling, P., et al. 2014. PTV margin definition in hypofractionated IGRT of localized prostate cancer using cone beam CT and orthogonal image pairs with fiducial markers. *Radiation oncology*. 9:229.

Ojerholm, E., Bekelman, J. E. 2018. Finding value for protons: The case of prostate cancer? *Seminars in radiation oncology.* 28(2):131–137.

Olberg, S., Green, O., Cai, B., et al. 2018. Optimization of treatment planning workflow and tumor coverage during daily adaptive magnetic resonance image guided radiation therapy (MR-IGRT) of pancreatic cancer. *Radiation oncology.* 13:51.

Oldham, M., Letourneau, D., Watt, L., et al. 2005. Cone-beam-CT guided radiation therapy: A model for on-line application. *Radiotherapy & oncology.* 75(3):271–278.

Olivera, G. H., Mackie, T. R. 2012. Megavoltage fan beam CT. In: Bourland, D. (ed.), *Image Guided Radiation Therapy.* Boca Raton, FL, USA: CRC Press, Taylor and Francis Group.

Ono, T., Miyabe, Y., Takahashi, K., et al. 2018. Geometric and dosimetric accuracy of dynamic tumor tracking during volumetric-modulated arc therapy using a gimbal mounted linac. *Radiotherapy & oncology.* 129:166–172.

Ortiz, L. P., Andreo, P., Cosset, J-M., et al. 2000. *ICRP Report No. 86: Prevention of accidents to patients undergoing radiation therapy.* Ottawa, Canada: ICRP.

Ortiz, L. P., Cossett, J. M., Dunscombe, P., et al. 2009. *ICRP Report No. 112: Preventing accidental exposures from new external beam radiation therapy technologies.* Ottawa, Canada: ICRP.

Ouzidane, M., Evans, J., Djemil, T. 2015. Image guidance and localization technologies for radiotherapy. In: Benedict, S. et al. (eds.), *Stereotactic Radiosurgery and Stereotactic Body Radiation Therapy.* London, UK: Routledge, Taylor and Francis Group.

Padhani, A. R., Khoo, V. S., Suckling, J. 1999. Evaluating the effect of rectal distension and rectal movement on prostate gland position using cine MRI. *International journal of radiation oncology, biology, physics.* 44(3):525–533.

Paganetti, H. 2014. Relative biological effectiveness (RBE) values for proton beam therapy. Variations as a function of biological endpoint, dose, and linear energy transfer. *Physics in medicine and biology.* 59:R419–R472.

Paliwal, B. R., DeLuca, P. M., Grein, E. E. et al. 2009. *Academic program recommendations for graduate degrees in medical physics: Report of the education and training of medical physicists committee.* College Park, MD, USA: American Association of Physicists in Medicine.

Pallotta, S., Vanzi, E., Simontacchi, G., et al. 2015. Surface imaging, portal imaging, and skin marker set-up vs. CBCT for radiotherapy of the thorax and pelvis. *Strahlentherapie Onkologie.* 191(9):726–733.

Pan, H., Cervino, L. I., Pawlicki, T. Frameless, real-time, surface imaging-guided radiosurgery: Clinical outcomes for brain metastases. *Neurosurgery-online.* 2012, 71(4). 844-852

Pantelis, E., Moutsatsos, A., Zourari, K. et al. 2012. On the output factor measurements of the Cyberknife iris collimator small fields. Experimental determination of the k correction factors for microchamber and diode detectors. *Medical physics.* 39:4875–4885.

Parodi, K. 2015. Vision 20/20 positron emission tomography in radiation therapy planning, delivery, and monitoring *Medical physics.* 42:7153.

Partouche, J., Chnura, S. J., Luke, J. J., et al. 2018. *ASTRO 2018 Abs.* Evaluation of a prototype treatment planning system (TPS) for biology-guided radiotherapy (BgRT) in the context of stereotactic body radiation therapy (SBRT) for oligo-metastases *International journal of radiation oncology, biology, physics.* 102(3):Supplement pages e514-e515.

Patel, I., Weston, S., Palmer, A. L., et al. (eds). 2018. *Physics aspects of quality control in radiotherapy: IPEM Report 81, 2nd Edition.* York, UK: Institute of Physics and Engineering in Medicine.

Pathmanathan, A., van As, N., Kerkmeijer, L., et al. 2018. Magnetic resonance imaging-guided adaptive radiation therapy: A "game changer" for prostate treatment? *International journal of radiation oncology, biology, physics.* 100(2):361–373.

Pawlicki, T., Coffey, M., Milosevic, M. 2017. Incident learning systems for radiation oncology: Development and value at the local, national and international level. *Clinical oncology (Royal College of Radiologists).* 29:562–567.

Pawlowski, J. M., Yang, E. S., Malcolm, A. W., et al. 2010. Reduction of dose delivered to organs at risk in prostate cancer patients via image-guided radiation therapy. *International journal of radiation oncology, biology, physics.* 76:924–934.

Pedroni, E., Meer, D., Bula, C., et al. 2011. Pencil beam characteristics of the next-generation proton scanning gantry of PSI: Design issues and initial commissioning results. *European physical journal plus.* 126:1–27.

Penn. 2017. *Penn Medicine treats world's first patient on Varian's Halcyon System.* 22nd September 2017. Available at: https://www.itnonline.com/content/penn-medicine-treats-worlds-first-patient-varians-halcyon-system (accessed on 3rd December 2018).

Petersen, R. P., Truong, P. T., Kader, H. A., et al. 2007. Target volume delineation for partial breast radiotherapy planning: Clinical characteristics associated with low interobserver concordance. *International journal of radiation oncology biology physics.* 69:41–48.

PHE (Public Health England). 2018a. *Safer radiotherapy: Radiotherapy newsletter of Public Health England.* September 2018, Issue 26. London, UK: PHE.

PHE (Public Health England). 2018b. *Safer radiotherapy: Radiotherapy newsletter of Public Health England: Supplementary Data Analysis Issue 26—full radiotherapy error data analysis April to July 2018.* London, UK: Public Health England.

PHE (Public Health England). 2018c. *Learning from the past 10 years of the Radiotherapy Clinical Site Visit.* London, UK: Public Health England.

Phillips, M. H., Singer, K., Miller, E., et al. 2000. Commissioning an image-guided localization system for radiotherapy. *International journal of radiation oncology, biology, physics.* 48(1):267–276.

Phillips, R., Da Silva, A., Radwan, N., et al. 2018. *ASTRO 2018 Abs.* PSMA-directed biologically-guided radiation therapy of castration-sensitive oligometastatic prostate cancer patients. *International journal of radiation oncology, biology, physics.* 102(3):Supplement pages S152.

Pidikiti, R., Patel, B., Maynard, M., et al. 2018. Commissioning of the world's first compact pencil-beam scanning proton therapy system. *Journal of applied clinical medical physics.* 19(1):94–105.

Podgorsak, E. (tech ed.) 2005. Quality assurance of external beam radiotherapy. In: *Radiation Oncology Physics: A Handbook for Teachers and Students.* Vienna, Austria: International Atomic Energy Agency.

Potter, R., Eriksen, J. G., Beavis, A. W., et al. 2012. Competencies in radiation oncology: A new approach for education and training of professionals for radiotherapy and oncology in Europe. *Radiotherapy & oncology.* 103:1–4.

Pouliot, J., Bani-Hashemi, A., Chen, J., et al. 2005. Low-dose megavoltage cone-beam CT for radiation therapy. *International journal of radiation oncology, biology, physics.* 61(2):552–560.

Praxiom website. *ISO 9000 2015: Plain English Definitions.* Available at: http://www.praxiom.com/iso-definition.htm#Quality_management (accessed on 3rd December 2018).

Price, R., Kim, J., Zheng, W., et al. 2016. Image guided radiation therapy using synthetic computed tomography images in brain cancer. *International journal of radiation oncology, biology, physics.* 95(4):1281–1289.

Qi, X. S., Hu, A. Y., Lee, S. P., et al. 2013. Assessment of interfraction patient setup for head-and-neck cancer intensity modulated radiation therapy using multiple computed tomography-based image guidance. *International journal of radiation oncology, biology, physics.* 86(3):432–439.

Queens. 2017. *Queen's Hospital first in the UK to transform cancer treatment with top of the range new radiotherapy machine.* 13th October 2017. Available at: https://www.bhrhospitals.nhs.uk/news/queens-hospital-first-in-the-uk-to-transform-cancer-treatment-with-top-of-the-range-new-radiotherapy-machine-1630 (Accessed on 3rd December 2018).

Quick, H. 2014. Integrated PET/MR. *Journal of magnetic resonance imaging.* 39:243–258.

Raaijmakers, A., Raaymakers, B., Lagendijk, J. 2008 Magnetic-field-induced dose effects in MR-guided radiotherapy systems: Dependence on the magnetic field strength. *Physics in medicine and biology.* 53:909–923.

Rajendran, R. R., Plastaras, J. P., Mick, R. 2010. Daily isocenter correction with electromagnetic-based localization improves target coverage and rectal sparing during prostate radiotherapy. *International journal of radiation oncology, biology, physics.* 76(4):1092–1099.

Ramey, S., Padgett, K., Lamichhane, N., et al. 2018. Dosimetric analysis of stereotactic body radiation therapy for pancreatic cancer using MR-guided Tri-60Co unit, MR-guided LINAC, and conventional LINAC-based plans *Practical radiation oncology.* 8(5):e312-e321.

Rankine, L. J., Mein, S., Cai, B., et al. 2017. Three-dimensional dosimetric validation of a magnetic resonance guided intensity modulated radiation therapy system. *International journal of radiation oncology, biology, physics.* 97:1095–1104.

Rasch, C., Barillot, I., Remeijer, P. et al. 1999. Definition of the prostate in CT and MRI: A multi-observer study. *International journal of radiation oncology, biology, physics.* 43(1):57–66.

RaySearch Laboratories website. *RayCare One Oncology Workflow: The next-generation OIS.* Available at: https://www.raysearchlabs.com/raycare/ (accessed on 3rd December 2018).

RCR (Royal College of Radiologists) website. Available at: https://www.rcr.ac.uk/(accessed on 3rd December 2018).

RCR (Royal College of Radiologists). 2008a. *On-target: Ensuring geometric accuracy in radiotherapy.* London, UK: RCR.

RCR (Royal College of Radiologists). 2008b. *Towards Safer Radiotherapy.* London, UK: RCR.

RCR (Royal College of Radiologists). 2008c *A guide to understanding the implications of the Ionising Radiation (Medical Exposure) Regulations in radiotherapy.* London, UK: RCR.

RCR (Royal College of Radiologists). 2013. *Guidance on the management and governance of additional radiotherapy capacity.* London, UK: RCR.

RCR (Royal College of Radiologists). 2016. *Clinical Oncology Syllabus.* Available at: https://www.rcr.ac.uk/sites/default/files/2016_curriculum_-_clinical_oncology_-_appendix_1_syllabus_15_november_2016.pdf (accessed on 3rd December 2018).

Reason, J. 1990. *Human Error.* Cambridge, UK: Cambridge University Press.

Reason, J. T. 2001. Understanding adverse event: The human factor. In: Vincent C. A. (ed), *Clinical Risk Management: Enhancing Patient Safety.* 2nd ed. London, UK: BMJ Publications.

RedJournal. *May 2016 Edition of International Journal of Radiation Oncology Biology Physics.* 95(1). Available at https://www.redjournal.org/ (accessed on 3rd December 2018).

RefleXion website. *RefleXion Technology 2018.* https://www.reflexion.com/technology/#1542263730702-4e0393bc-e0b0 (accessed on 3rd December 2018).

Reichner, C. A., Collins, B. T., Gagnon, G. J., et al. 2005. The placement of gold fiducials for CyberKnife stereotactic radiosurgery using a modified transbronchial needle aspiration technique. *Journal of bronchology.* 12:193–195.

Reiner, B., Bownes, P., Buckley, D. L., et al. 2017. Quantifying the effects of positional uncertainties and estimating margins for Gamma-Knife® fractionated radiosurgery of large brain metastases. *Journal of neurosurgery. SBRT* 4(4):275–287.

Richter, C., Pausch, G., Barczyk, S. 2016. First clinical application of a prompt gamma based in vivo proton range verification system. *Radiotherapy & oncology.* 118:232–237.

Riley, C., Cox, C., Graham, S., et al. 2018. Varian Halcyon dosimetric comparison for multiarc VMAT prostate and head-and-neck cancers. *Medical dosimetry.* (In press)

Rit, S., Nijkamp, J., van Herk, M., et al. 2011. Comparative study of respiratory motion correction techniques in cone-beam computed tomography. *Radiotherapy & oncology.* 100(3):356–359.

Robinson, D., Liu, D., Steciw, S. 2012. An evaluation of the Clarity 3-D ultrasound system for prostate localization. *Journal of applied clinical medical physics.* 13(4):100–112.

Rogus, R., Stern, R., Kubo, H. 1999. Accuracy of a photogrammetry-based patient positioning and monitoring system for radiation therapy. *Medical physics*. 26(5):721–728.

Romesser, P. B., Cahlon, O., Scher, E., et al. 2016. Proton beam radiation therapy results in significantly reduced toxicity compared with intensity-modulated radiation therapy for head and neck tumors that require ipsilateral radiation. *Radiotherapy & oncology*. 118:286–292.

Rong, Y., Welsh, J. S. 2011. Dosimetric and clinical review of helical tomotherapy. *Expert review of anticancer therapy*. 11(2):309–320.

Royal Australian and New Zealand College of Radiologists (RANZCR). *New revised radiation oncology practice standards*. 2018. Available at: https://www.ranzcr.com/whats-on/news-media/240-new-revised-radiation-oncology-practice-standards (accessed on 3rd December 2018).

Rozendaal, R., Mijnheer, B., van Herk, M., et al. 2014. In vivo portal dosimetry for head-and-neck VMAT and lung IMRT: Linking c-analysis with differences in dose–volume histograms of the PTV. *Radiotherapy & oncology*. 112:396–401.

Ruschin, M., Komljenovic, P. T., Ansell, S., et al. 2013. Cone beam computed tomography image guidance system for a dedicated intracranial radiosurgery treatment unit. *International journal of radiation oncology, biology, physics*. 85(1):243–250.

Ruschin, M., Nayebi, N., Carlsson, P., et al. 2010. Performance of a novel repositioning head frame for Gamma knife Perfexion and image-guided linac-based intracranial stereotactic radiotherapy. *International journal of radiation oncology, biology, physics*. 78:306–313.

Rybovic, M., Halkett, G. K., Banati, R. B. 2008. Radiation therapists' perceptions of the minimum level of experience required to perform portal image analysis. *Radiography*. 14:294–300.

Samei, E., Ikejimba, L. C., Harrawood, B. P., et al. 2018. Report of AAPM Task Group 162: Software for planar image quality metrology. *Medical physics*. 45(2):e32–e39.

Santos, D., Aubin, J., Fallone, B., et al. 2012. Magnetic shielding investigation for a 6 MV in-line linac within the parallel configuration of a linac-MR system. *Medical physics*. 39:788–797.

Sayeh, S., Wang, J., Main, W. T., et al. 2007. Respiratory motion tracking for robotic radiosurgery. In: Urschel, H. C. et al. (eds.), *Robotic Radiosurgery: Treating Tumors that Move with Respiration*. Berlin, Germany: Springer-Verlag.

Scally, G., Donaldson, L. J. 1998. Clinical governance and the drive for quality improvement in the new NHS in England. *BMJ*. 317:61–65.

Scarborough, T. J., Golden, N. M., Ting, J. Y., et al. 2006. Comparison of ultrasound and implanted seed marker prostate localization methods: Implications for image-guided radiotherapy. *International journal of radiation oncology, biology, physics*. 65(2):378–387.

Scheuermann, R., Kennedy, C., Mihailidis, D., et al. 2017. Performance study of a prototype straight-through linac delivery system with an EPID assembly. *Radiotherapy & oncology*. 123(Suppl 1):S506–S507.

Schippers, J., Lomax, A., Garonna, A., et al. 2018. Can technological improvements reduce the cost of proton radiation therapy? *Seminars in radiation oncology*. 28:150–159.

Schlosser, J., Gong, R., Bruder, R., et al. 2016. Robotic intrafractional US guidance for liver SABR: System design, beam avoidance, and clinical imaging. *Medical physics*. 43(11):5951–5963.

Schoffel, P., Harms, W., Sroka-Perez, G., et al. 2007. Accuracy of a commercial optical 3-D surface imaging system for realignment of patients for radiotherapy of the thorax. *Physics in medicine & biology*. 52:3949–3963.

Schultheiss, T. E., Wong, J., Liu, A., et al. 2007. Image-guided total marrow and total lymphatic irradiation using helical tomotherapy. *International journal of radiation oncology, biology, physics*. 67(4):1259–1267.

Schwartz, D. L., Garden, A. S., Thomas, J., et al. 2012. Adaptive radiotherapy for head-and-neck cancer: Initial clinical outcomes from a prospective trial. *International journal of radiation oncology, biology, physics*. 83:986–993.

Schwarz, M., Cattaneo, G. M., Marrazzo, L. 2017. Geometrical and dosimetrical uncertainties in hypofractionated radiotherapy of the lung: A review. *Physica medica*. 36:126–139.

Seco, J., Spadea, M. F. 2015. Imaging in particle therapy: State of the art and future perspective. *Acta oncologica*. 54:1254–1258.

Sen, A., West M. K. 2009. Commissioning experience and quality assurance of helical tomotherapy machines. *Journal of medical physics*. 34(4):194–199.

Seppenwoolde, Y., Berbeco, R. I., Nishioka, S., et al. 2007. Accuracy of tumor motion compensation algorithm from a robotic respiratory tracking system: A simulation study. *Medical physics*. 34:2774–2784.

Serago, C., Chungbin, S., Buskirk, S., et al. 2002. Initial experience with ultrasound localization for positioning prostate cancer patients for external beam radiotherapy. *International journal of radiation oncology biology physics*. 53(5):1130–1138.

Seyedin, S., Mawlawi, O. R., Turner, L. M., et al. 2015. Use of emission guided radiation therapy can better spare critical structures compared with intensity modulated radiation therapy, volumetric modulated arc therapy, or proton therapy. *International journal of radiation oncology, biology, physics*. 93(3) (Supplement): E612.

Shafiq, J., Barton, M., Noble, D., et al. 2009. An international review of patient safety measures in radiotherapy practice. *Radiotherapy & oncology*. 92:15–21.

Shah, A., Aird, E., Shekhdar, J. 2012. Contribution to normal tissue dose from concomitant radiation for two common kV-CBCT systems and one MVCT system used in radiotherapy. *Radiotherapy & oncology*. 105(1):139–144.

Shah, A., Dvorak, T., Curry, M. S., et al. 2013a. Clinical evaluation of interfractional variations for whole breast radiotherapy using 3-dimensional surface imaging. *Practical radiation oncology*. 3(1):16–25.

Shah, A., Kupelian, P., Waghorn, B., et al. 2013b. Real-time tumor tracking in the lung using an electromagnetic tracking system. *International journal of radiation oncology biology physics*. 86(3):477–483.

Shah, A. P., Langen, K. M., Ruchala, K. J., et al. 2008. Patient dose from megavoltage computed tomography imaging. *International journal of radiation oncology, biology, physics*. 70(5):1579–1587.

Shakeshaft, J., Perez, M., Tremethick, L., et al. 2014. ACPSEM ROSG Oncology-PACS and OIS working group recommendations for quality assurance. *Australasian physical & engineering sciences in medicine*. 37:3–13.

Sharpe, M. B., Velec, M., Brock, K. K. 2012. Image registration and segmentation in radiation therapy. In: Bourland, D. (ed.), *Image Guided Radiation Therapy*. Boca Raton, FL, USA: CRC Press, Taylor and Francis Group.

Shirato, H., Seppenwoolde, Y., Kitamura, K., et al. 2004. Intrafractional tumor motion: Lung and liver. *Seminars in radiation oncology*. 14:10–18.

Sihono, D., Vogel, L., Weiss, C., et al. 2017. A 4-D ultrasound real-time tracking system for external beam radiotherapy of upper abdominal lesions under breath-hold. *Strahlentherapie onkologie*. 193(3):213–220.

Skarsgard, D., Cadman, P., El-Gayed, A., et al. 2010. Planning target volume margins for prostate radiotherapy using daily electronic portal imaging and implanted fiducial markers, *Radiation oncology*. 5:52.

Society and College of Radiographers (SCoR). 2013. *Education and Career Framework for the Radiography workforce*. London, UK: Society of Radiographers.

Society and College of Radiographers professional document. 2013. *Image guided radiotherapy (IGRT) clinical support programme in England, 2012-2013*. London, UK: SCoR.

Soete, G., Verellen, D., Michielsen, D., et al. 2002. Clinical use of stereoscopic x-ray positioning of patients treated with conformal radiotherapy for prostate cancer. *International journal of radiation oncology biology physics*. 54(3):948–952.

Solberg, T. D., Medin, P. M., Ramirez, E., et al. 2014. Commissioning and initial stereotactic ablative radiotherapy experience with Vero. *Journal of applied clinical medical physics*. 15(2):205–225.

Sonke, J. J. 2010. *Adaptive radiation therapy*. Available at: https://www.avl.nl/media/517544/Adaptive%20radiotherapy%20(ART).pdf (accessed on 3rd December 2018).

Sonni, I., Iagaru, A. 2016. PET imaging toward individualized management of urologic and gynecologic malignancies. *PET clinics*. 11:261–272.

Sovik, A., Malinen, E., Olsen, D. 2010. Adapting biological feedback in radiotherapy. *Seminars in radiation oncology*. 20:138–146.

Spreeuw, H., Rozendaal, R., Olaciregui-Ruiz, I., et al. 2016. Online 3-D EPID-based dose verification: Proof of concept. *Medical physics*. 43:3969–3974.

SRO. April 2018. *April 2018 edition of seminars in radiation oncology (28(2), 2018)*. Available at https://www.sciencedirect.com/journal/seminars-in-radiation-oncology/vol/28/issue/2

St. Aubin, J., Steciw, S., Fallone, B. 2010. Waveguide detuning caused by transverse magnetic fields on a simulated in-line 6 MV linac. *Medical physics*. 37(9):4751–4754.

Stam, M. K., van Vulpen, M., Barendrecht, M. M. 2013. Kidney motion during free breathing and breathhold for MR-guided radiotherapy. *Physics in medicine and biology*. 58(7):2235–2245.

StandardsStores, 9000 Store website. *ISO 9001 Quality Management System—what is a QMS?* Available at: https://the9000store.com/iso-9001-2015-requirements/what-is-iso-9001-quality-management-system/ (accessed on 3rd December 2018).

Stewart, J., Lim, K., Kelly, V., et al. 2010. Automated weekly replanning for intensity-modulated radiotherapy of cervix cancer. *International journal of radiation oncology, biology, physics*. 78:350–358.

Stieler, F., Wenz, F., Abo-Madyan, Y., et al. 2016. Adaptive fractionated stereotactic Gamma Knife radiotherapy of meningioma using integrated stereotactic cone-beam-CT and adaptive re-planning (a-gkFSRT). *Strahlentherapie und onkologie*. 192:815–819.

Stroom, J. C., De Boer, H. C. J., Huizenga, H., Visser, A. G. 1999. Inclusion of geometrical uncertainties in radiotherapy treatment planning by means of coverage probability. *International journal of radiation oncology biology physics*. 43(4):905–919.

Stroom, J. C., Heijmen, B. J. M. 2002 .Geometrical uncertainties, radiotherapy planning margins, and the ICRU-62 report. *Radiotherapy & oncology*. 64:75–83.

Su, Z., Zhang, L., Murphy, M., et al. 2011. Analysis of prostate patient setup and tracking data: Potential intervention strategies. *International journal of radiation oncology, biology, physics*. 81(3):880–887.

Sullivan, A., Ding, G. 2015. Additional imaging guidance dose to patient organs resulting from x-ray tubes used in CyberKnife image guidance system. *Medical physics*. 42:3264.

Sykes, J. R., Amer, A., Czajka, J., et al. 2005. A feasibility study for image guided radiotherapy using low dose, high speed, cone beam x-ray volumetric imaging. *Radiotherapy & oncology*. 77(1):45–52.

Symonds, P., Deehan, C., Mills, J. A., et al. (eds.). 2012. *Walter and Miller's Textbook of Radiotherapy: Radiation Physics, Therapy and Oncology*. London, UK: Elsevier Churchill Livingstone.

Szyszko, T. A., Cook, G. J. R. 2018. PET/CT and PET/MRI in head and neck malignancy. *Clinical radiology*. 73:60–69.

Tabacchi, E., Fanti, S., Nanni, C. 2016. The possible role of PET imaging toward individualized management of bone and soft tissue malignancies. *PET clinics*. 11:285–296.

Takahashi, Y., Verneris, M. R., Dusenbery K. E., et al. 2013. Peripheral dose heterogeneity due to the thread effect in total marrow irradiation with helical tomotherapy. *International journal of radiation oncology, biology, physics*. 87(4):832–839.

Takayama, K., Mizowaki, T., Kokubo, M., et al. 2009. Initial validations for pursuing irradiation using a gimbals tracking system. *Radiotherapy & oncology*. 93(1):45–49.

Tanaka, H., Hayashi, S., Ohtakara, K., et al. 2011. Usefulness of CT-MRI fusion in radiotherapy planning for localized prostate cancer. *Journal of radiation research*. 52:782–788.

Tanyi, J., He, T., Summers, P., et al. 2010. Assessment of planning target volume margins for intensity-modulated radiotherapy of the prostate gland: Role of daily inter- and intrafraction motion. *International journal of radiation oncology biology physics*. 78(5):1579–1585.

Tenn, S., Solberg, T., Medin, P. 2005. Targeting accuracy of an image guided gating system for stereotactic body radiotherapy. *Physics in medicine & biology*. 50(23):5443–5462.

The British Institute of Radiology. 2003. *Geometric uncertainties in radiotherapy: Defining the planning target volume*. London: BIR.

The British Standards Institution. ISO 9001 Quality Management resources web page. Available at: https://www.bsigroup.com/en-GB/iso-9001-quality-management/Resources-for-ISO-9001/ (accessed on 3rd December 2018).

The Society of Radiographers website. Available at: https://www.sor.org/(accessed on 3rd December 2018).

Thilmann, C., Nill, S., Tucking, T., et al. 2006. Correction of patient positioning errors based on in-line cone beam CTs: Clinical implementation and first experiences. *Radiation Oncology*. 1:16.

Thomas, G. A. 2004. Solid cancers after therapeutic radiation – can we predict which patients are most at risk? *Clinical oncology journal (Royal College of Radiologists)*. 16(6):429–434.

Thomas, S. J., Vinall, A., Poynter, A., Routsis, D. 2010. A multicenter timing study of intensity-modulated radiotherapy planning and delivery. *Clinical oncology*. 22:658–665.

Thornqvist,. S, Hysing, L., Tuomikoski, L., et al. 2016. Adaptive radiotherapy strategies for pelvic tumors—a systematic review of clinical implementations. *Acta oncologica*. 55(8):943–958.

Thorwarth, D., Alber, M. 2010. Implementation of hypoxia imaging into treatment planning and delivery. *Radiotherapy & oncology*. 97:172–175.

Torheim, T., Malinen, E., Hole, K., et al. 2017. Autodelineation of cervical cancers using multiparametric magnetic resonance imaging and machine learning. *Acta oncologica*. 56(6):806–812.

Torresin, A., Brambillai, M., Monti, A., et al. 2015. Review of potential improvements using MRI in the radiotherapy workflow. *Zeitschrift für Medizinische Physik*. 25(3):210–220.

Trebeschi, S., van Griethuysen, J., Lambregts, D., et al. 2017. Deep learning for fully-automated localization and segmentation of rectal cancer on multiparametric MR. *Scientific reports*. 7(1):5301.

Uh, J., Krasin, M., Li, Y., et al. 2017. Quantification of pediatric abdominal organ motion with a 4-dimensional magnetic resonance imaging method. *International journal of radiation oncology, biology, physics*. 99(1):227–237.

Uh, S., Stewart, J., Moseley, J., et al. 2014. Hybrid adaptive radiotherapy with on-line MRI in cervix cancer IMRT. *Radiotherapy & oncology*. 110(2):323–328.

Unkelbach, J., Pagenetti, H., 2018. Robust proton treatment planning: Physical and biological optimization. *Seminars in radiation oncology*. 28(2):88–96.

Valk, P., Delbeke, D., Bailey, D. 2006. *Positron Emission Tomography—Clinical Practice*. London: Springer-Verlag.

Van den Begin, R., Engels, B., Boussaer, M., et al. 2016. Motion management during SBRT for oligometastatic cancer: Results of a prospective phase II trial. *Radiotherapy & oncology*. 119:519–524.

Van den Bosch, M., Ollers, M., Reyman, B., et al. 2017. Automatic selection of lung cancer patients for adaptive radiotherapy using cone-beam CT imaging. *Physics and imaging in radiation oncology*. 1:21–27.

Van den Brekel, M., Stel, H., Castelijns, J., et al. 1990. Cervical lymph node metastasis: Assessment of radiologic criteria. *Radiology*. 177:379–384.

Van der Meer, S., Bloemen-van Gurp, E., Hermans, J., et al. 2013. Critical assessment of intra-modality 3-D ultrasound imaging for prostate IGRT compared to fiducial markers. *Medical physics*. 40(7):071707-1-071707-11.

Van der Merwe, D., Van Dyk, J., Healy, B., et al. 2017. Accuracy requirements and uncertainties in radiotherapy: A report of the internal atomic energy agency. *Acta oncologica.* 56(1):1–6.

Van Elmpt, W., De Ruysscher, D., van der Salm, A. 2012. The PET-boost randomised phase II dose-escalation trial in non-small cell lung cancer. *Radiotherapy & oncology.* 104:67–71.

Van Elmpt, W., McDermott, L., Nijsten, S., et al. 2008. A literature review of electronic portal imaging for radiotherapy dosimetry. *Radiotherapy & oncology.* 88:289–309.

Van Gelder, R., Wong, S., Le, A., et al. 2018. Experience with an abdominal compression band for radiotherapy of upper abdominal tumours. *Journal of medical radiation sciences.* 65:48–54.

Van Heijst, T., Hartogh, M., Lagendijk, J., et al. 2013. MR-guided breast radiotherapy: Feasibility and magnetic-field impact on skin dose. *Physics in medicine and biology.* 58:5917–5930.

Van Herk, M. 2004. Errors and margins in radiotherapy. *Seminars in radiation oncology.* 14(1):52–64.

Van Herk, M. 2007. Different styles of image-guided radiotherapy. *Seminars in radiation oncology.* 17:258–267.

Van Herk, M., McWilliam, A., Dubec, M., et al. 2018. Magnetic resonance imaging-guided radiation therapy: A short strengths, weaknesses, opportunities, and threats analysis. *International journal of radiation oncology, biology, physics.* 101(5):1057–1060.

Van Herk, M., Remeijer, P., Rasch, C., Lebesque, J. V. 2000. The probability of correct target dosage: Dose population histograms for deriving treatment margins in radiotherapy. *International journal of radiation oncology biology physics.* 47(4):1121–1135.

Van Nunen, A., Van der Sangan, M. J. C., Van Boxtel, M., et al. 2017. Cone-beam CT-based position verification for oesophageal cancer: Evaluation of registration methods and anatomical changes during radiotherapy. *Technical innovations & patient support in radiation oncology.* 3–4:30–36.

Van Sornsen De Koste, J. R., De Boer, H. C. J., Schuchhard-Schipper, R. H., Senan, S., et al. 2003. Procedures for high precision setup verification and correction of lung cancer patients using CT-simulation and digitally reconstructed radiographs (DRR). *International journal of radiation oncology, biology, physics.* 55(3):804–810.

Van Wickle, J., Paulson, E., Landry, J., et al. 2017. Adaptive radiation dose escalation in rectal adenocarcinoma: A review. *Journal of gastrointestinal oncology.* 8(5):902–914.

Vargas, C., Saito, A., Hsi, W., et al. 2010. Cine-magnetic resonance imaging assessment of intrafraction motion for prostate cancer patients supine or prone with and without a rectal balloon. *American journal of clinical oncology.* 33(1):11–16.

Varian website. 2018a. *Halcyon.* Available at: https://www.varian.com/oncology/products/treatment-delivery/halcyon (accessed on 3rd December 2018).

Varian website. 2018b. *ARIA OIS for radiation oncology.* Available at: https://www.varian.com/en-gb/oncology/products/software/information-systems/aria-ois-radiation-oncology (accessed on 3rd December 2018).

Vestergaard, A., Hafeez, S., Muren, L., et al. 2016. The potential of MRI-guided online adaptive re-optimisation in radiotherapy of urinary bladder cancer. *Radiotherapy & oncology.* 118:154–159.

Waddington, S. P., McKenzie, A. L. 2004. Assessment of effective dose from concomitant exposures required in verification of the target volume in radiotherapy. *British journal of radiology.* 77(919):557–561.

Walker, G. V., Johnson, J., Edwards, T., et al. 2015. Factors associated with radiation therapy incidents in a large academic institution. *Practical radiation oncology.* 5:21–27.

Walston, S., Quick, A., Kuhn, K., et al. 2017. Dosimetric considerations in respiratory-gated deep inspiration breath-hold for left breast irradiation. *Technology in cancer research & treatment.* 16(1):22–32.

Wan, S., Stillwaugh, L., Prichard, H., et al. 2008. Evaluation of a 3D ultrasound system for image guided radiation therapy for prostate cancer. *International journal of radiation oncology biology physics.* 72(1):S555.

Wang, A., Paysan, P., Brehm, M, et al. 2016. Advanced scatter correction and iterative reconstruction for improved cone-beam CT imaging on the TrueBeam Radiotherapy Machine [White paper]. AAPM Annual Meeting. DOI: 10.1118/1.4957761.

Wang, H., Garden, A. S., Zhang, L., et al. 2008. Performance evaluation of automatic anatomy segmentation algorithm on repeat or four-dimensional computed tomography images using deformable image registration method. *International journal of radiation oncology, biology, physics.* 72:210–219.

Wang, L., Solberg, T., Medin, P., et al. 2001. Infrared patient positioning for stereotactic radiosurgery of extracranial tumors. *Computers in biology and medicine.* 31(2):101–111.

Warrington, J. 2006. Commissioning for stereotaxis. Chap. 13 of Kirby, M. C., et al. (eds.), *Acceptance testing and commissioning of linear accelerators: IPEM Report 94.* York, UK: Institute of Physics and Engineering in Medicine.

Webster, G., Kilgallon, J., Rowbottom, C., et al. 2009. A novel imaging technique for fusion of high-quality immobilized MR images of the head and neck with CT scans for radiotherapy target delineation. *British journal of radiology.* 82:497–503.

Wen, N., Guan, H., Hammoud, R., et al. 2007. Dose delivered from Varian's CBCT to patients receiving IMRT for prostate cancer. *Physics in medicine and biology.* 52:2267–2276.

Westerley, D. C., Schefter, T. E., Kavanagh, B. D., et al. 2012. High-dose MVCT image guidance for stereotactic body radiation therapy. *Medical physics.* 39(8):4812–4819.

Weygand, J., Fuller, C., Ibbott, G., et al. 2016. Spatial precision in magnetic resonance imaging-guided radiation therapy: The role of geometric distortion. *International journal of radiation oncology, biology, physics.* 95(4):1304–1316.

Wiant, D., Squire, S., Liu, H., et al. 2016. A prospective evaluation of open face masks for head and neck radiation therapy. *Practical radiation oncology.* 6:e259–e267.

Wiant, D., Verchick, Q., Gates, P., et al. 2016. A novel method for radiotherapy patient identification using surface imaging. *Journal of applied clinical medical physics.* 17(2):271–278.

Willoughby, T., Kupelian, P., Pouliot, J., et al. 2006. Target localization and real-time tracking using the Calypso 4-D localization system in patients with localized prostate cancer. *International journal of radiation oncology biology physics.* 65(2):528–534.

Willoughby, T., Lehman, J., Bencomo, J. A., et al. 2012. Quality assurance for nonradiographic radiotherapy localization and positioning systems: Report of Task Group 147. *Medical physics.* 39(4):1728–1747.

Witte, M. G., Sonke, J-J., Siebers, J. O. D., van Herk, M. 2017. Beyond the margin recipe: The probability of correct target dosage and tumor control in the presence of a dose limiting structure. *Physics in medicine and biology.* 62:7874–7888.

Wojcieszynski, A., Hill, P., Rosenberg, S., et al. 2017. Dosimetric comparison of real-time MRI-guided tri-Cobalt-60 versus linear accelerator-based stereotactic body radiation therapy lung cancer plans. *Technology in cancer research & treatment.* 16:366–372.

Wojcieszynski, A. P., Rosenberg, S. A., Brower, J. V. 2016. Gadoxetate for direct tumor therapy and tracking with real-time MRI-guided stereotactic body radiation therapy of the liver. *Radiotherapy & oncology.* 118(2):416–418.

Woodford, C., Yartsev, S., Dar, A., et al. 2007. Adaptive radiotherapy planning on decreasing gross tumor volumes as seen on megavoltage computed tomography images. *International journal of radiation oncology, biology, physics.* 69(4):1316–1322.

World Health Organization (WHO). 1988. *Quality assurance in radiotherapy.* Geneva, Switzerland: WHO.

World Health Organization (WHO). 2008. *Radiotherapy Risk Profile: Technical Manual.* Geneva, Switzerland: WHO.

Wright, G., Harrold, N., Hatfield, P., et al. 2017. Validity of the use of nose tip motion as a surrogate for intracranial motion in mask-fixated frameless Gamma Knife® Icon™ therapy. *Journal of radiosurgery and SBRT*. 4:289–301.

Wu, Q. J., Thongphiew, D., Wang, Z., et al. 2008. On-line re-optimization of prostate IMRT plans for adaptive radiation therapy. *Physics in medicine and biology*. 53:673–691.

Wu, W. C., Leung, W. S., Kay, S. S., et al. 2011. A comparison between electronic portal imaging device and cone beam CT in radiotherapy verification of nasopharyngeal carcinoma. *Medical dosimetry*. 36(1):109–112.

Xie, Y., Bentefour, E. H., Janssens, G., et al. 2017. Prompt gamma imaging for in vivo range verification of pencil beam scanning proton therapy. *International journal of radiation oncology, biology, physics*. 99:210–218.

Yan, D. 2010. Adaptive radiotherapy: Merging principle into clinical practice. *Seminars in radiation oncology*. 20(2):79–83.

Yan, D., Vicini, F., Wong, J., et al. 2002. Adaptive radiation therapy. *Physics in medicine and biology*. 42:123–132.

Yan, H., Yin, F., Kim, J. 2003. A phantom study on the positioning accuracy of the Novalis Body system. *Medical physics*. 30(12):6052–6060.

Yang, J., Yamamoto, T., Mazin, S., et al. 2014. The potential of positron emission tomography for intratreatment dynamic lung tumor tracking: A phantom study. *Medical Physics*. 41(2):021718-1–14.

Yang, J., Yamamoto, T., Thielemans, K., et al. 2011. A feasibility study for real-time tumour tracking using positron emission tomography (PET). *Medical physics*. SU-E-J–156: https://doi.org/10.1118/1.3611924 (accessed on 3rd December 2018).

Yartsev, S., Bauman, G. 2016. Target margins in radiotherapy of prostate cancer. *British journal of radiology*. 89(1067):20160312.

Yin, F.-F., Wong, J., Balter, J., et al. 2009. *The role of in-room kV x-ray imaging for patient setup and target localization: Report of task group 104*. College Park, MD, USA: American Association of Physicists in Medicine.

Yoon, M., Shen, D. H., Kim J., et al. 2011. Craniospinal irradiation techniques: A dosimetric comparison of proton beams with standard and advanced photon radiotherapy. *International journal of radiation oncology, biology, physics*. 81(3):637–646.

Yu, A., Fahimian, B., Million, L. 2017. A robust and affordable table indexing approach for multi-isocentre dosimetrically matched fields. *Cureus*. 9(5):e1270.

Yu, H., Caldwell, C., Balogh, J., et al. 2014. Toward magnetic resonance-only simulation: Segmentation of bone in MR for radiation therapy verification of the head. *International journal of radiation oncology, biology, physics*. 89(3):649–657.

Yu, P., Gandhidasan, S., Miller, A. A. 2010. Different usage of the same oncology information system in two hospitals in Sydney—lessons go beyond the initial introduction. *International journal of medical informatics*. 79:422–429.

Zagar, T., Kaidar-Person, O., Tang, X., et al. 2017. Utility of deep inspiration breath hold for left-sided breast radiation therapy in preventing early cardiac perfusion defects: A prospective study. *International journal of radiation oncology biology physics*. 97(5):903–909.

Zeverino, M., Agostinelli, S., Taccini, G., et al. 2012. Advances in the implementation of helical tomotherapy-based total marrow irradiation with a novel field junction technique. *Medical dosimetry*. 37:314–320.

Zhao, H., Wang, B., Sarkar, V., et al. 2016. Comparison of surface matching and target matching for image-guided pelvic radiation therapy for both supine and prone patient positions. *Journal of applied clinical medical physics*. 17(3):14–24.

Zheng, Y., Sun, X., Wang, J., et al. 2014. FDG-PET/CT imaging for tumor staging and definition of tumor volumes in radiation treatment planning in non-small cell lung cancer. *Oncology letters*. 7:1015–1020.

Zhou, D., Quan, H., Yan, D., et al. 2018. A feasibility study of intrafractional tumor motion estimation based on 4-D CBCT using diaphragm as surrogate. *Journal of applied clinical medical physics.* 19(5):525–531.

Zhu, X., El Fakhri, G., 2013. Proton therapy verification with PET imaging. *Theranostics.* 3(10):731–740.

Zou, W., Dong, L., Teo, B-K. K. 2018. Current state of image guidance in radiation oncology: Implications for PTV margin expansion and adaptive therapy. *Seminars in radiation oncology.* 28:238–247.

Index